Interacting Gravitational, Electromagnetic, Neutrino and Other Waves

In the Context of
Einstein's General Theory of Relativity

Interacting Gravitational, Electromagnetic, Neutrino and Other Waves

In the Context of Einstein's General Theory of Relativity

Anzhong Wang

Baylor University, USA

World Scientific

NEW JERSEY · LONDON · SINGAPORE · BEIJING · SHANGHAI · HONG KONG · TAIPEI · CHENNAI · TOKYO

Published by

World Scientific Publishing Co. Pte. Ltd.

5 Toh Tuck Link, Singapore 596224

USA office: 27 Warren Street, Suite 401-402, Hackensack, NJ 07601

UK office: 57 Shelton Street, Covent Garden, London WC2H 9HE

British Library Cataloguing-in-Publication Data
A catalogue record for this book is available from the British Library.

INTERACTING GRAVITATIONAL, ELECTROMAGNETIC, NEUTRINO AND OTHER WAVES
In the Context of Einstein's General Theory of Relativity

ISBN 978-981-121-148-5 (hardcover)
ISBN 978-981-121-149-2 (ebook for institutions)
ISBN 978-981-121-150-8 (ebook for individuals)

For any available supplementary material, please visit
https://www.worldscientific.com/worldscibooks/10.1142/11585#t=suppl

Desk Editor: Ng Kah Fee

Typeset by Stallion Press
Email: enquiries@stallionpress.com

Printed in Singapore

To My Wife and Son

Preface

Einstein's theory of general relativity is a classical theory of space, time and gravitation. In competition with other gravitational theories, it is the one best supported by experiments and observations. This is particularly true, after the detection of the first gravitational wave by the Laser Interferometer Gravitational-Wave Observatory in September 14, 2015 (Abbott *et al.*, 2016), a century-old prediction of Albert Einstein (1916) by using his by-then new theory of relativity (Einstein, 1915a, 1915b). Soon after, ten more gravitational waves were detected from binary mergers of not only two stellar-mass black holes (Abbott *et al.*, 2019) but also two neutron stars (Abbott *et al.*, 2017), jointly observed by the Virgo Collaboration after August 2017. The studies of Einstein's theory in the strong field regime have been further promoted by the direct detection of the supermassive black hole, M87, in April 2017 by the Event Horizon Telescope, but with a final and processed image released only on April 10, 2019 (Akiyama *et al.*, 2019).

However, over the last several decades, the complexity and the nonlinear nature of the field equations of the theory have constantly hindered us from deeply understanding its content and richness. As a result, most of such studies have been numerical (Baumgarte and Shapiro, 2010), once the strong field regime is involved.

Nevertheless, the implication of Einstein's theory of general relativity (GR) cannot be completely exploited without analytical studies of the Einstein field equations. Such analytical investigations usually follow two different paths. One is to find approximate solutions, generally based on some linear approximations. Such linearized field equations can be well justified and represent good approximations to the real world, such as the cosmological linear perturbations (Lyth and Liddle, 2009), and the problem of a binary compact system in its inspiral phase (Maggiore, 2008). But, it is also true that such obtained solutions commonly lose the nonlinear features of the field equations, and many interesting properties, such as the occurrence of spacetime singularities, are due to the nonlinearity of the field equations. The second path is to find exact solutions, which can remedy the shortage of the former, but again, because of the complexity of the field equations, exact solutions often have high degree of symmetry, which would not be present in the real world. Therefore, one must always take cautions when studying exact solutions and distinguish properties due to the nonlinearity of the field equations from those due to the assumed symmetries.

With the above in mind, the search for exact solutions has already achieved great success in the stationary axisymmetric spacetimes (Stephani *et al.*, 2009), as a result of the revolutionary new techniques based on the recent mathematical developments in the theory of nonlinear differential equations (Gardner *et al.*, 1967). Meanwhile, there have been also successful attempts to employ these same techniques in the studies of spacetimes with two commuting space-like Killing vectors, including the ones with plane or cylindrical symmetry (Griffiths and Podolský, 2009; Bronnikov, Santos and Wang, 2019).

In this book, we consider only the spacetimes for nonlinearly interacting gravitational plane waves, a subject that has been extensively studied since the pioneering works of Szekeres (1970, 1972) and Khan and Penrose (1971), and well documented by Griffiths (1991, and republished in 2016). Among the differences between Griffiths' book and our current one, two aspects are worth mentioning here. First, in this book, applying the distribution theory to the spacetimes of colliding plane waves, a technique first introduced to general relativity by Taub (1980), we develop a complete description of the Einstein field equations in terms of the Newman–Penrose formalism (Newman and Penrose, 1962, 1963). Second, in such spacetimes the frame, which is normally adopted to define the polarization of the interacting gravitational plane waves, is usually rotating along each of the two colliding wave paths. As a result, defining polarization with respect to this kind of bases losses its meaning when we compare the polarization of a gravitational wave at different locations and times. To overcome this problem, we first introduce a parallelly-transported frame along the gravitational wave path, and then project its polarization to this parallelly transported frame (Wang, 1991b), so that such defined polarization angle has a unique physical interpretation, and is independent of the choice of coordinates, that is, if such a defined angle changes in one set of coordinates, it will change in any of others, or vice versa. Once this is made clear, we turn to the gravitational analog of the electromagnetic Faraday rotation (1846a, 1846b) for colliding gravitational plane waves. Using the Bianchi identities, we show explicitly that the changes of the polarization of a plane gravitational wave is exactly due to the nonlinear interaction with the oppositely moving gravitational and/or matter waves (Wang, 1991b). Without such nonlinear interactions, the polarization of each of the gravitational plane waves remains constant. This has been further shown by studying exactly solutions for the spacetimes of colliding plane waves.

Among such solutions are those that describe both colliding purely gravitational plane waves and colliding gravitational plane waves coupled with matter fields. The collision and interaction of such waves are interesting in several aspects. One of them is that the nonlinearity of general relativity shows up explicitly in the failure of the principle of superposition. As a result, colliding plane gravitational waves generically develop spacetime singularities in the future of the collision. Killing–Cauchy horizons can be also developed in the interacting region of the two colliding plane waves in some particular cases (Chandrasekhar and Xanthopoulos, 1986a), but they are not stable (Yutsever, 1987; Griffiths, 2005), and with small but generic perturbations they ought to be turned into spacetime singularities (Clarke and Hayward, 1989).

One may argue that such formations of spacetime singularities are due to the high symmetry of the spacetime. In particular, with plane symmetry the incoming waves always have infinitely large amount of energy. Therefore, in more realistic cases the singularities might be replaced by high curved regions. However, recently numerical studies of two colliding plane-fronted massless particle waves in asymptotically flat spacetimes showed that spacetime singularities are still formed, even the total energy of the incoming waves is finite, but now they are hidden inside horizons, that is, now black holes are formed (Pretorius and East, 2018).

The structure of this book is as follows. The introduction for the fundamental concepts and material are arranged so that they are contained in two chapters, Chapters 1 and 2. Specifically, Chapter 1 includes the introduction of some basic physical quantities, the Newman–Penrose formalism, the physical interpretation of the optical scalars, and the Einstein field equations in terms of distribution-valued tensors. On the other hand, in Chapter 2, the definition of a single plane gravitational wave is given, and its polarization angle and amplitude are given explicitly in terms of the

Weyl scalars. The spacetime singularities are also studied, and found that spacetimes of gravitational plane waves are generally always singular when written in the Baldwin, Jeffery and Rosen coordinates (Baldwin and Jeffery, 1926; Rosen, 1937), and are free of singularities only in some particular cases (Wang *et al.*, 2018). The relevance of such singularities to memory effects (see, for example, Favata, 2010; Bieri, Garfinkle and Yunes, 2017) and soft theorems of gravitons (Hawking, Perry and Strominger, 2016, 2017; Strominger, 2017) is also mentioned.

In Chapter 3, after the Weyl and Ricci scalars, as well as the Bianchi identities, are written in terms of distributions for the spacetimes of colliding gravitational plane waves, the polarization of a gravitational plane wave in the interaction region is defined with respect to a parallelly-transported frame along its wave path, whereby a gravitational analog of the electromagnetic Faraday rotation is investigated for various types of collisions. The nature of the singularities formed in the interaction region of the two colliding plane gravitational wave due to their mutual focus, and some methods for generating exact solutions are also discussed.

Chapters 4 is devoted to the studies of the collision and subsequent interaction of two pure gravitational plane waves, while Chapter 5 is devoted to the studies of a gravitational plane wave colliding with a matter wave. The latter can be an impulsive and/or a shock dust shell, an electromagnetic wave, or a neutrino wave. In these chapters, we present three classes of exact solutions of the Einstein field equations, which include most of the known diagonal and non-diagonal solutions found so far for the collision of gravitational plane waves without or with the presence of matter fields. The main properties of these solutions and the effects of polarization of the colliding gravitational plane waves on the formation and nature of singularities are investigated in detail.

In Chapter 6, we study isometries between the internal spacetimes of black holes and the interaction regions of two colliding plane waves, and show that the interiors of all the known black holes have a one-to-one correspondence to the interaction regions of colliding plane waves. These include the Schwarzschild, Reissner–Nordström, Kerr–Neman–NUT solutions, and the ones with a cosmological constant. Finally, in Chapter 7, we present our concluding remarks, and point several directions one can pursue in the future, including some observations.

This book is based in a large part on the author's Ph.D. dissertation (Wang, 1991f), which was written about three decades ago. But, the materials and topics are still relevant to current studies of gravitational waves, especially the phenomena due to the nonlinear effects of the Einstein field equations, such as the formation and nature of spacetime singularities due to the mutual focus of the two interacting gravitational waves, and the gravitational analog of the electromagnetic Faraday rotation. Another important topic is the memory effect (Zeldovitch and Polnarev, 1974), which is not considered in this book, but can be easily extended to such studies by following the introduction of Section 2.5 of Chapter 2. Memory effects have been extensively studied recently (Favata, 2010; Bieri, Garfinkle and Yunes, 2017), as the current generation of gravitational wave detectors might be able to observe such nonlinear phenomena (Lasky *et al.*, 2016).

Finally, taking this opportunity, I would like to express my gratitude to my teachers, colleagues and friends, who are so important to me, not only because of the publication of this book, but also because of the way that they change and sharp my life. Certainly, first I would like to express my gratitude to Prof. Dimitri G. Tsoubelis, my Ph.D. supervisor, for leading me to the field that this small book covers. It is difficult to imagine that without his constant interest and cooperation the present book would have been completed. I am greatly indebted to Profs. Basills C. Xanthopoulos and Charalampos Kolassis for giving me the benefit of consultation and helpful discussions. My deep thanks go to Prof. Pericles Tsekeris and his family members for constant support and encouragements. My thanks also go to Profs. S. Evangelou, P. Kanti, K. Tamvakis, V. Tsikoudi and I. Vergados for their encouragements and help. The discussions with Dr. Athanasios Economou on soliton techniques were invaluable, and I appreciated so much his help and friendship. I also thank Mrs. Kaity Stoghianidis and Dr. John Rizos for their useful discussions and help. I owe my thanks to Dr. Anna van Gogh for both her help and friendship.

I have also benefited greatly from numerous discussions with my collaborators, Profs. E. Abdalla, J.S. Alcaniz, K. Bamba, M.L. Bedran, G.A. Benesh, K.A. Bronnikov, Ronggen Cai, Roberto Chan, Gerald Cleaver, M.F. da Silva, J.C.N. de Araujo, A. De Felice, H.P. de Oliveira, Chikun Ding, X.-J. Fang, J. Gariel, O. Goldoni, J. Greenwald, A.E. Gumrukcuoglu, L. Herrera, E.W. Hirschmann, P.R. Holvorcem, M. Ishak, Mubashar Jamil, J.L. Jing, Klaus Kirsten, D. Langlois, Patricio S. Letelier, J. Lenells, Zhong-Heng Li, Kai Lin, Hai-Shan Liu, Tan Liu, Guo-Liang Lu, JianXin Lu, Roy Maartens, Shinji Mukohyama, K. Noui, R. Opher, Supriya Pan, A. Papazoglou, P. Rocha, M. Sami, N.O. Santos, H.J. Schmidt, Qin Sheng, Fu-Wen Shu, Parampreet Singh, J. Soda, A.F.F. Teixeira, M. Tian, David Wand, Bin Wang, Zhao-Jun Wang, H. Wei, Zhong-Chao Wu, Jie Yang, W.-Q. Yang, Shao-Jun Zhang, X. Zhang, Wen Zhao, and Tao Zhu, and my former and current postdocs and Ph.D. students, Drs. Madhurima Bhattacharjee,

Jakub Bilski, Carlos Frederico Brandt, Jaime F. da Rocha, Michael Devin, Antonino Flachi, Jaijai Geng, Yungui Gong, Fei-Hung He, Yongqing Huang, Bao-Fei Li, Alexandre Yasuda Miguelote, Gil de Oliveira Neto, Jacob Oost, Hamed Pejhan, Paulo R. Pereira, Filipe de Moraes Paiva, V.H. Satheeshku-mar, Mohd Shahalam, Manabendra Sharma, Andreas Tziolas, José A.C. Nogales Vera, Mew-Bing Wan, Xinwen Wang, Qiang Wu, and Messrs Jared R. Fier, Wen-Cong Gan, Baowen Li, Chao Zhang and Xiang Zhao. I am also very grateful and thankful to my classmates and lifetime friends, Profs. Zhaojun Wang and Yannan Yang, for their constant encouragement and support.

In the preparation of this book, I have been greatly assisted by a number of colleagues. In particular, I am extremely grateful to Mr. Kah Fee Ng, Senior Editor from World Scientific Publishing Company, for the most helpful comments and assistance.

I cannot express adequately my indebtedness to my wife, Yumei, for her invaluable assistance with the manuscript of this book, as well as for her time, patience, encouragement, inspiration and unconditional support.

Finally, to those people whose contribution I can no longer quite recall I offer both my thanks and apology.

Anzhong Wang

China Spring, TX
September, 2019

Contents

List of Figures

List of Figures

Chapter 1

Fundamentals of Einstein's Theory of General Relativity

In this chapter, we shall provide some fundamentals that are to be used in this book. Among these are the definitions of some basic physical quantities (for example, the Riemann tensor, parallel transport, geodesic deviation, and the optical scalars), the Newman–Penrose (NP) formalism, and the Einstein field equations with distribution-valued tensors. For more details on Einstein's theory of general relativity, we refer the readers to the textbooks of Wald (1984), D'Inverno (2003), and Carroll (2004). For the NP formalism, we refer the readers to Newman and Penrose (1962, 1963), Frolov (1979), Kramer $et\ al.$ (1980), Griffiths (1991), and Stephani $et\ al.$ (2009), while for generalized functions, to Gelfand and Shilov (1964), Challifour (1972), Vladimirov (1979), Zemanian (1987), and Ding and Ding (2005).

In this book, we shall adopt the following conventions: the metric signature is $(+, -, \ldots, -)$, and the Christoffel symbols, the Riemann, Ricci and Einstein tensors, and the Ricci scalar are defined, respectively, by

$$\Gamma^a_{bc} = \frac{1}{2} g^{ad} \left(g_{dc,b} + g_{bd,c} - g_{bc,d} \right),$$

$$R^a{}_{bcd} = \Gamma^a_{bd,c} - \Gamma^a_{bc,d} + \Gamma^a_{ec}\Gamma^e_{bd} - \Gamma^a_{ed}\Gamma^e_{bc},$$

$$R_{ab} = R^c{}_{acb}, \quad R = g^{ab} R_{ab}, \quad G_{ab} = R_{ab} - \frac{1}{2} R g_{ab},$$

where

$$g_{ab,c} \equiv \frac{\partial g_{ab}}{\partial x^c}, \quad R^c{}_{acb} \equiv \sum_{c=0}^{N-1} R^c{}_{acb},$$

etc., with N denoting the dimension of the spacetime.[1] Then, the Einstein field equations take the form,

$$R_{ab} - \frac{1}{2}Rg_{ab} - \Lambda g_{ab} = \varkappa T_{ab},$$

where Λ is the cosmological constant, T_{ab} the energy–momentum tensor, and $\varkappa \equiv 8\pi G/c^4$, with G denoting the Newtonian constant. In addition, we also define

$$X_{(ab)} \equiv \frac{1}{2}\left(X_{ab} + X_{ba}\right), \quad X_{[ab]} \equiv \frac{1}{2}\left(X_{ab} - X_{ba}\right).$$

1.1. Spacetime Manifolds

From the mathematical point of view, the fundamental object of Einstein's theory of general relativity (GR) is the four-dimensional (4D) spacetime manifold $(\Omega, g_{\mu\nu})$, where Ω is a connected 4D Hausdoff C^∞ manifold,[2] and $g_{\mu\nu}$ is a symmetric Lorentz metric tensor, or simply a metric on Ω. For the study of differential geometry, we refer the readers to Farnsworth, Fink, Porter and Thompson (1972) and Westenholz (1981). The points in Ω are labeled by a general coordinate system (x^0, x^1, x^2, x^3), often written as x^μ ($\mu = 0, 1, 2, 3$). We use the convention that Greek indices take the values $0, 1, 2, 3$, and repeated Greek indices are to be summed over these values unless specified otherwise.

According to the principle of covariance, all coordinate systems are equivalent for the description of physical phenomena. Thus, the choice of coordinate systems is arbitrary. If we go from one coordinate system, say, x^μ, to another, say, x'^μ, a contravariant vector y^μ and a covariant vector y_μ transform, respectively, as

$$y'^\mu = \frac{\partial x'^\mu}{\partial x^\nu}y^\nu, \quad y'_\mu = \frac{\partial x^\nu}{\partial x'^\mu}y_\nu, \tag{1.1}$$

and a mixed tensor such as $y^\mu{}_{\nu\lambda}$ as

$$y'^\mu{}_{\nu\lambda} = \frac{\partial x'^\mu}{\partial x^\sigma}\frac{\partial x^\rho}{\partial x'^\nu}\frac{\partial x^\delta}{\partial x'^\lambda}y^\sigma{}_{\rho\delta}, \tag{1.2}$$

etc.

---

[1]In this book, we mainly focus on $(3+1)$-dimensional spacetimes, although many formulas developed here can be easily extended to spacetimes with any (higher) dimension.
[2]A quantity F is said to be C^n if and only if all derivatives of F up to the nth order exist and are continuous (Hawking and Ellis, 1973).

The contravariant tensor $g^{\mu\nu}$ in terms of $g_{\mu\nu}$ is defined by

$$g^{\mu\nu}g_{\mu\lambda} = \delta^\nu_\lambda, \tag{1.3}$$

where δ^μ_ν is the Kronecker delta, which is unity for $\mu = \nu$ (no summation is taken) and zero otherwise. By using $g^{\mu\nu}$ and $g_{\mu\nu}$, we can raise and lower the indices as

$$y^\mu \equiv g^{\mu\nu}y_\nu, \quad y_\mu \equiv g_{\mu\nu}y^\nu. \tag{1.4}$$

We regard tensors derived by such raising and lowering of indices as representing the same quantity, since by raising an index and subsequently lowering it we recover the original tensor.

All the local information about the spacetime is contained in the metric $g_{\mu\nu}$, which determines the square of the spacetime interval ds between two infinitesimally separated events or points x^μ and $x^\mu + dx^\mu$ as

$$ds^2 = g_{\mu\nu}dx^\mu dx^\nu. \tag{1.5}$$

The contravariant vector dx^μ is said to be time-like, space-like or null according to whether ds^2 is positive, negative or zero, respectively. The spacetime manifold Ω has three space-like and one time-like dimensions.

Since the Einstein field equations contain the second derivatives of the metric, while the Bianchi identities contain its third derivatives, it is necessary to require $g_{\mu\nu}$ to be at least C^3 and $x^\mu = x^\mu(x'^\mu)$ to be at least C^4, so that the Einstein field equations are defined everywhere, while the Bianchi identities are defined at every point of the spacetime manifold. However, as we shall show in Section 1.9, these conditions can be relaxed in the sense of distributions (Challifour, 1972; Vladimirov, 1979; Zemanian, 1987; Ding and Ding, 2005).

1.2.   Covariant Differentiations, the Riemann Tensor and Einstein's Field Equations

To generalize the ordinary (partial) differentiation to the Riemannian geometry, one introduces an additional structure into the manifold, an affine connection, ∇, which assigns to each vector field X ($\equiv X^\mu\partial_\mu$) on Ω a differential operator, ∇_X, which maps an arbitrary vector field Y into a vector field $\nabla_x Y$.

Associated with each metric, we can endow the manifold with a unique torsion-free connection by requiring

$$\nabla g = 0. \tag{1.6}$$

In a local coordinate basis $\{\partial_\lambda\}$, Eq. (1.6) can be written in the following form:

$$\nabla_{\partial_\lambda} g_{\mu\nu} = g_{\mu\nu,\lambda} - g_{\mu\delta}\Gamma^\delta_{\nu\lambda} - g_{\delta\nu}\Gamma^\delta_{\mu\lambda} = 0, \tag{1.7}$$

where $\partial_\lambda \equiv \partial/\partial x^\lambda$, a comma "," denotes the partial differentiation with respect to the indicated argument, and $\Gamma^\lambda_{\mu\nu}(=\Gamma^\lambda_{\nu\mu})$ are called the Christoffel connection coefficients (symbols), and the connection itself called the Christoffel connection. From Eq. (1.7) and by using the symmetry of $\Gamma^\lambda_{\mu\nu}$, we find

$$\Gamma^\lambda_{\mu\nu} = \frac{1}{2}g^{\lambda\delta}\left(g_{\mu\delta,\nu} + g_{\nu\delta,\mu} - g_{\mu\nu,\delta}\right). \tag{1.8}$$

The covariant differentiation for a contravariant or covariant vector is defined as

$$A^\mu_{\;;\nu} = A^\mu_{\;,\nu} + \Gamma^\mu_{\nu\lambda}A^\lambda, \quad A_{\mu;\nu} = A_{\mu,\nu} - \Gamma^\lambda_{\mu\nu}A_\lambda, \tag{1.9}$$

and for a mixed tensor such as $A^\mu_{\;\nu\lambda}$ as

$$A^\mu_{\;\nu\lambda;\sigma} = A^\mu_{\;\nu\lambda,\sigma} + \Gamma^\mu_{\delta\sigma}A^\delta_{\;\nu\lambda} - \Gamma^\delta_{\nu\sigma}A^\mu_{\;\delta\lambda} - \Gamma^\delta_{\lambda\sigma}A^\mu_{\;\nu\delta}, \tag{1.10}$$

and so on, where a semicolon ";" denotes the covariant differentiation.

Under a coordinate transformation, say, from x^μ to x'^μ, the connection coefficients, $\Gamma^\lambda_{\mu\nu}$, transform as

$$\Gamma'^\mu_{\nu\lambda} = \frac{\partial x'^\mu}{\partial x^\rho}\frac{\partial x^\sigma}{\partial x'^\nu}\frac{\partial x^\delta}{\partial x'^\lambda}\Gamma^\rho_{\sigma\delta} + \frac{\partial^2 x^\sigma}{\partial x'^\nu \partial x'^\lambda}\frac{\partial x'^\mu}{\partial x^\sigma}. \tag{1.11}$$

Therefore, $\Gamma^\mu_{\alpha\beta}$ is not a tensor. For a covariant vector A_μ, it can be shown that

$$A_{\mu;\nu;\lambda} - A_{\mu;\lambda;\nu} = A_\delta R^\delta_{\;\mu\nu\lambda}, \tag{1.12}$$

where $R^\delta_{\;\mu\nu\lambda}$ is the Riemann tensor defined by

$$R^\sigma_{\;\mu\nu\lambda} \equiv \Gamma^\sigma_{\mu\lambda,\nu} - \Gamma^\sigma_{\mu\nu,\lambda} + \Gamma^\delta_{\mu\lambda}\Gamma^\sigma_{\delta\nu} - \Gamma^\delta_{\mu\nu}\Gamma^\sigma_{\delta\lambda}, \tag{1.13}$$

and Eq. (1.12) is called the Ricci identity.

The Riemann tensor has the symmetry,

$$R_{\sigma\mu\nu\lambda} = -R_{\mu\sigma\nu\lambda} = -R_{\sigma\mu\lambda\nu}, \tag{1.14a}$$

$$R_{\sigma\mu\nu\lambda} = R_{\nu\lambda\sigma\mu}, \tag{1.14b}$$

$$R_{\sigma\mu\nu\lambda} + R_{\sigma\lambda\mu\nu} + R_{\sigma\nu\lambda\mu} = 0, \tag{1.14c}$$

and satisfies the Bianchi identities

$$R^\sigma{}_{\mu\nu\lambda;\rho} + R^\sigma{}_{\mu\rho\nu;\lambda} + R^\sigma{}_{\mu\lambda\rho;\nu} = 0. \tag{1.15}$$

The Ricci tensor $R_{\mu\lambda}$ is defined by

$$R_{\mu\lambda} \equiv g^{\sigma\nu} R_{\sigma\mu\nu\lambda} = R^\delta{}_{\mu\delta\lambda}. \tag{1.16}$$

From Eqs. (1.14a)–(1.14c) and (1.16), it is easy to show that

$$R_{\mu\lambda} = R_{\lambda\mu}. \tag{1.17}$$

The Ricci scalar is defined by

$$R \equiv g^{\sigma\lambda} R_{\sigma\lambda} = R^\lambda{}_\lambda. \tag{1.18}$$

By contracting the Bianchi identities on the pairs of indices $\mu\nu$ and $\sigma\rho$, we find that

$$\left(R_{\mu\nu} - \frac{1}{2} g_{\mu\nu} R \right)_{;\lambda} g^{\lambda\nu} = 0. \tag{1.19}$$

The tensor $G_{\mu\nu} \equiv R_{\mu\nu} - (1/2) g_{\mu\nu} R$ is called the Einstein tensor.

We are now in a position to write down the Einstein field equations, which are the fundamental differential equations of GR,

$$R_{\mu\nu} - \frac{1}{2} g_{\mu\nu} R - \Lambda g_{\mu\nu} = \varkappa T_{\mu\nu}, \tag{1.20}$$

where $\varkappa$ $(\equiv 8\pi G/c^4)$ is the Einstein coupling constant, and Λ the cosmological constant. The tensor $T_{\mu\nu}$ denotes the energy–stress tensor of the source producing the gravitational field. Without loss of generality, one can always choose units such that $\varkappa = 1$.

It must be noted, however, that the form of the Einstein field equations used by Chandrasekhar (1983) and Pirani (1964) is not consistent with the requirement that the energy density of matter fields must be positive (see, for example, Weyl, 1922; Lichnerowicz, 1955; Throne, 1967; Landau and Lifshitz, 1972; Misner, Thorne and Wheeler, 1973; Wald, 1984; Islam, 1985; D'Inverno, 2003; Carroll, 2004).

The combination of Eqs. (1.19) and (1.20) gives

$$T^{\mu\nu}{}_{;\nu} = 0, \tag{1.21}$$

which are the equations for the conservation of energy and momentum (stress) of the source.

1.3. Curves, Parallel Transportation and Geodesics

A curve in a Riemannian space is defined by points $x^\mu(\lambda)$ where x^μ are suitably differentiable functions of the real parameter λ, varying over some interval of the real line. The curve is time-like, space-like, or null according to whether its tangent vector $(dx^\mu/d\lambda)$ is time-like, space-like or null.

In Euclidean geometry, for an arbitrary vector field X we will say that X is parallelly transported along the curve if $X^\mu{}_{,\nu}(dx^\nu/d\lambda) = 0$. In a general differentiable manifold with a connection, we define analogously that a vector X is parallelly transported along the curve if its covariant derivative $X^\mu{}_{;\nu}(dx^\nu/d\lambda)$ along this curve is zero, that is, if

$$X^\mu{}_{;\nu}\frac{dx^\nu}{d\lambda} = \left(X^\mu{}_{,\nu} + \Gamma^\mu_{\nu\delta}X^\delta\right)\frac{dx^\nu}{d\lambda} = \frac{dX^\mu}{d\lambda} + \Gamma^\mu_{\nu\delta}X^\delta\frac{dx^\nu}{d\lambda} = 0. \quad (1.22)$$

A similar definition holds for tensors. Given any curve $x^\mu(\lambda)$ with the end points $\lambda = \lambda_1$ and $\lambda = \lambda_2$, the theory of solutions of ordinary differential equations shows that if $\Gamma^\lambda_{\mu\nu}$ are suitably differentiable functions of x^μ, we obtain a unique tensor at $\lambda = \lambda_2$ by parallelly transporting it from the point $\lambda = \lambda_1$, along the curve, to the point $\lambda = \lambda_2$ (Hawking and Ellis, 1973).

A particular case is the covariant derivative of the tangent vector itself along the curve $x^\mu(\lambda)$. The curve is said to be a geodesic, if the tangent vector is parallelly transported along this curve, i.e., if

$$\frac{d^2x^\mu}{d\lambda^2} + \Gamma^\mu_{\nu\delta}\frac{dx^\nu}{d\lambda}\frac{dx^\delta}{d\lambda} = 0. \quad (1.23)$$

When the equation for a geodesic is reduced to the form of Eq. (1.23), we say that it is affinely parameterized. It should be noted that such defined affine parameter λ is not unique, and still subjected to the rescaling and shift of origin: $\tilde{\lambda} = \alpha\lambda + \lambda_0$, where α and λ_0 are constant. Equation (1.23) also describes the motion of a free particle.

1.4. Geodesic Deviations

A major problem that has to be solved in the study of gravitational radiation is how to identify a gravitational radiation field. The problem arises because of the principle of the equivalence, which says that the motion of a test particle in a gravitational field is independent of its mass and composition. This implies that mechanical phenomena are the same in an accelerated laboratory as in the earth's gravitational field, if observations are confined to a region over which the variation in the earth's gravitational field is observationally small. Thus, in a local experiment we cannot distinguish an inertial field from a genuine gravitational one. However, if we are allowed to carry out non-local experiments, we can distinguish one from another by observing the variation of the field rather than the field itself. In GR, this variation is described by the Riemann tensor which specifies the relative acceleration of neighboring free particles.

Let us consider a one-parameter family of geodesics $\Gamma(w)$ specified by the equations

$$x^\mu = x^\mu(\lambda, w), \tag{1.24}$$

where we assume x^μ to be at least twice continuously differentiable functions of both λ and w. The parameter w varies from one geodesic to another while λ varies along each of geodesics. For fixed w, we have the geodesic equations [see Eq. (1.23)],

$$\frac{\partial^2 x^\mu}{\partial \lambda^2} = -\Gamma^\mu_{\nu\delta} \frac{\partial x^\nu}{\partial \lambda} \frac{\partial x^\delta}{\partial \lambda}, \quad x^\mu = x^\mu(\lambda, w). \tag{1.25}$$

We might, in general, identify λ, with the arc length on each of the geodesics. We prefer, however, to leave λ to be defined just by Eq. (1.25) so that our following discussions remain also valid for null geodesics.

The family of geodesics gives rise to the vector fields

$$t^\mu(\lambda, w) = \frac{\partial x^\mu(\lambda, w)}{\partial \lambda}, \tag{1.26a}$$

$$\eta^\mu(\lambda, w) = \frac{\partial x^\mu(\lambda, w)}{\partial w}, \tag{1.26b}$$

where $t^\mu(\lambda, w)$ is the tangent vector along each geodesic, and $\eta^\mu(\lambda, w)$ is the vector that describes the deviation of two points on two infinitesimally near geodesics, which have the same parameter value λ. The vector η^μ is usually called the geodesic deviation vector.

From Eq. (1.26b) we find that the covariant differentiation of η^μ along each geodesic is given by

$$\frac{D\eta^\mu}{D\lambda} \equiv \eta^\mu{}_{;\nu} \frac{\partial x^\nu}{\partial \lambda} = \frac{\partial \eta^\mu}{\partial \lambda} + \Gamma^\mu_{\nu\delta}\eta^\delta \frac{\partial x^\nu}{\partial \lambda} = \frac{\partial t^\mu}{\partial w} + \Gamma^\mu_{\nu\delta}\eta^\delta t^\nu. \tag{1.27}$$

The remarkable fact is that the second differentiation of η^μ will bring us directly to the Riemann tensor. Actually, we have

$$\frac{D^2\eta^\mu}{D\lambda^2} = \frac{\partial}{\partial\lambda}\left(\frac{D\eta^\mu}{D\lambda}\right) + \Gamma^\mu_{\nu\delta}t^\nu\frac{D\eta^\delta}{D\lambda}$$

$$= \frac{\partial}{\partial w}\left(\frac{\partial^2 x^\mu}{\partial\lambda^2}\right) + \Gamma^\mu_{\nu\delta,\rho}t^\rho t^\nu\eta^\delta + \Gamma^\mu_{\nu\delta}\eta^\delta\frac{\partial^2 x^\nu}{\partial\lambda^2}$$

$$+ \Gamma^\mu_{\nu\delta}t^\nu\frac{\partial^2 x^\delta}{\partial\lambda\partial w} + \Gamma^\mu_{\nu\delta}t^\nu\frac{\partial^2 x^\delta}{\partial\lambda\partial w} + \Gamma^\mu_{\nu\delta}\Gamma^\delta_{\rho\sigma}t^\nu t^\rho\eta^\sigma. \qquad (1.28)$$

By Substituting Eq. (1.25) into Eq. (1.28), we find the well-known geodesic deviation equations

$$\frac{D^2\eta^\mu}{D\lambda^2} = -R^\mu{}_{\nu\delta\sigma}\eta^\delta t^\nu t^\sigma = R^\mu{}_{\nu\delta\sigma}t^\nu t^\delta\eta^\sigma, \qquad (1.29)$$

where $R^\mu{}_{\nu\lambda\sigma}$ is the Riemann tensor given by Eq. (1.13).

To illustrate the physical meaning of the geodesic deviation equations, let us consider a time-like geodesic, say, C. We introduce an orthogonal triad of space-like vectors, $\lambda^\mu_{(a)}$ ($a = 1, 2, 3$). Throughout this book, we use the convention that the indices inside parentheses denote tetrad indices, Roman indices take the values $1, 2, 3$, and repeated Roman indices are to be summed over these values, unless some specific statement to the contrary is made. These space-like vectors are assumed orthogonal to each other and to the tangent vector $\lambda^\mu_{(0)} = t^\mu$,

$$\lambda^\mu_{(\alpha)}\lambda^\nu_{(\beta)}g_{\mu\nu} = \lambda^\mu_{(\alpha)}\lambda_{\mu(\beta)} = \eta_{\alpha\beta}, \qquad (1.30)$$

where $\eta_{\alpha\beta}$ denotes the Minkowski metric, given by

$$[\eta_{\alpha\beta}] = \begin{bmatrix} 1 & 0 & 0 & 0 \\ 0 & -1 & 0 & 0 \\ 0 & 0 & -1 & 0 \\ 0 & 0 & 0 & -1 \end{bmatrix}. \qquad (1.31)$$

The tangent vector $\lambda^\mu_{(0)}$ can be interpreted physically as the four-velocity of an observer whose worldline is C, and the space-like vectors $\lambda^\mu(a)$ as rectangular coordinate axes used by this observer. For the sake of convenience, we assume that the orientations of the axes are fixed so that they are not rotating as determined by local dynamical experiments (see, for example, Pirani, 1964). This means that the vectors $\lambda^\mu_{(a)}$ are parallelly transported along C,

$$\lambda^\mu_{(a);\nu}t^\nu = 0. \qquad (1.32)$$

Without loss of generality, we also assume that η^μ is orthogonal to $\lambda_{(0)}^\mu$. Thus, the tetrad components of the deviation vector η^μ are

$$\eta^{(a)} = \eta^{(a)(\sigma)}\lambda_{(\sigma)}^\mu \eta^\nu g_{\mu\nu} = \lambda_\nu^{(a)}\eta^\nu, \quad \eta^{(0)} = 0, \tag{1.33}$$

where $\lambda_\nu^{(a)} \equiv \eta^{(a)(\sigma)}\lambda_{(\sigma)}^\mu g_{\mu\nu}$, $\eta^{(\mu)(\nu)} \equiv \eta^{\mu\nu}$. The components of η^a represent the spatial coordinates of a particle that moves nearby the observer, who moves along the geodesic C [sec Fig. 1.1].

Contracting Eq. (1.29) with $\lambda_\mu^{(a)}$ and using Eq. (1.32), we find that the acceleration of the particle relative to the observer is given by

$$\frac{d^2\eta^{(a)}}{d\tau^2} = -K^{(a)(b)}\eta_{(b)}, \tag{1.34}$$

where

$$K^{(a)(b)} \equiv R_{\mu\nu\rho\sigma}t^\nu t^\sigma \lambda^{\mu(a)}\lambda^{\rho(b)}, \tag{1.35}$$

are some of the tetrad components of the Riemann tensor. In writing Eq. (1.34), we replaced the parameter λ by the proper time τ measured by the observer using his own clock.

On the other hand, let us consider the same question in the framework of Newtonian gravitational theory. To be distinguishable, we use t as the time used by the observer and $\zeta^\mu(t)$ as the coordinate position of the particle relative to the observer. The gravitational field is described by the Newtonian potential ϕ. If $\zeta^\mu(t)$ is infinitesimal, then the equation of motion for

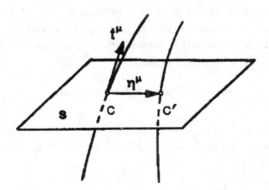

Fig. 1.1. The hypersurface S is space-like, spanned by the three space-like vectors, $\lambda_{(a)}^\mu$ $(a = 1, 2, 3)$.

the observer and the particle are given, respectively, by

$$\frac{d^2 x^a}{dt^2} = -\partial^a \phi,$$

$$\frac{d^2 x^a}{dt^2} + \frac{d^2 \zeta^a}{dt^2} = -\left\{\partial^a \phi\right\}|_{x+\zeta} = -\partial^a \phi - \zeta^b \partial^a \partial^b \phi, \tag{1.36}$$

where the derivatives of ϕ are evaluated at the point x^μ. It then follows that

$$\frac{d^2 \zeta^a}{dt^2} = -K^{ab} \zeta^b, \quad K^{ab} \equiv \partial^a \partial^b \phi. \tag{1.37}$$

The condition for the Laplacian potential $\nabla^2 \phi = 0$ leads to

$$K^{aa} = 0. \tag{1.38}$$

The similarity between Eqs. (1.37) and (1.34) is evident. Moreover, we even have $K^{(a)(a)} = 0$, whenever the Einstein vacuum field equations are satisfied.

The above consideration provides additional support for the choice of the field equation

$$R_{\mu\nu} = 0, \tag{1.39}$$

as a description of a free gravitational field.

1.5. Decompositions of the Riemann Tensor

The Riemann tensor $R^\mu{}_{\nu\lambda\rho}$ defined by Eq. (1.13) has 20 independent components, whereas the Ricci tensor $R_{\mu\nu}$ defined by Eq. (1.16) has only 10. Physically, it is convenient to decompose the Riemann tensor into three parts, which are all irreducible representations of the full Lorentz group (Kramer *et al.*, 1980; Stephani *et al.*, 2009),

$$R_{\mu\nu\lambda\rho} = C_{\mu\nu\lambda\rho} + E_{\mu\nu\lambda\rho} + G_{\mu\nu\lambda\rho}, \tag{1.40}$$

where

$$E_{\mu\nu\lambda\rho} \equiv \frac{1}{2} \left(g_{\mu\lambda} S_{\nu\rho} + g_{\nu\rho} S_{\mu\lambda} - g_{\nu\lambda} S_{\mu\rho} - g_{\mu\rho} S_{\nu\lambda} \right),$$

$$G_{\mu\nu\lambda\rho} \equiv \frac{1}{12} \left(g_{\nu\rho} g_{\mu\lambda} - g_{\nu\lambda} g_{\mu\rho} \right) R, \tag{1.41}$$

$$S_{\mu\nu} \equiv R_{\mu\nu} - \frac{1}{4} g_{\mu\nu} R.$$

In Eq. (1.41), the tensor $S_{\mu\nu}$ denotes the traceless part of the Ricci tensor. The Weyl tensor $C_{\mu\nu\lambda\rho}$ is thought of as representing the free gravitational field (Szekeres, 1965), and has all the symmetries of the Riemann tensor [see Eqs. (1.14a)–(1.14c)]. In addition, it is also traceless

$$C^{\lambda}{}_{\mu\lambda\nu} = 0. \tag{1.42}$$

Combining the fact that the Weyl tensor has all the symmetries of the Riemann tensor and Eq. (1.42), we can see that the Weyl tensor has only 10 independent components. At any given point of the spacetime, these components are completely independent of the Ricci tensor. Globally, however, the Weyl and Ricci tensors are not independent, as they are connected by the Bianchi identities [see Eq. (1.15)]. These identities can be now written in the form (Kundt and Trumper, 1962; Szekeres, 1966)

$$C_{\mu\nu\sigma\rho;}{}^{\rho} = R_{\sigma[\mu;\nu]} - \frac{1}{6}g_{\sigma[\mu}R_{;\nu]}, \tag{1.43}$$

where the square bracket denotes the antisymmetrization,

$$A_{[\mu\nu]} \equiv \frac{1}{2}\left(A_{\mu\nu} - A_{\nu\mu}\right). \tag{1.44}$$

The remarkable analogy between the Bianchi identities of Eq. (1.43) and the Maxwell equations,

$$F^{\mu\nu}{}_{;\nu} = j^{\mu}, \tag{1.45}$$

suggests that the Bianchi identities represent the interaction between the free gravitational field and matter. If we define the tensor $J_{\mu\nu\sigma}$ as

$$J_{\mu\nu\sigma} \equiv R_{\sigma[\mu;\nu]} - \frac{1}{6}g_{\sigma[\mu}R_{,\nu]}, \tag{1.46}$$

we have

$$J_{\mu\nu\lambda;}{}^{\lambda} = 0, \tag{1.47}$$

which strongly resembles the equation for the conservation of charges in electrodynamics

$$J^{\lambda}{}_{;\lambda} = 0. \tag{1.48}$$

Hence, $J_{\mu\nu\lambda}$ defined by Eq. (1.46) can be considered as representing a matter current, which consists of the parts of the source that interact with the free gravitational field. These parts are called gravitationally active, while the parts of the source that do not contribute to $J_{\mu\nu\lambda}$ are called gravitationally inert. The propagation of the free gravitational field is independent of the inert parts of the source.

An equivalent form for the decompositions of Eqs. (1.40) and (1.41) is given by

$$R_{\mu\nu\lambda\rho} = C_{\mu\nu\lambda\rho} + \frac{1}{2} \left[g_{\mu\lambda} R_{\nu\rho} + g_{\nu\rho} R_{\mu\lambda} - g_{\nu\lambda} R_{\mu\rho} - g_{\mu\rho} R_{\nu\lambda} \right]$$
$$+ \frac{1}{6} \left[g_{\mu\rho} g_{\nu\lambda} - g_{\mu\lambda} g_{\nu\rho} \right] R. \tag{1.49}$$

When the Weyl tensor $C_{\mu\nu\lambda\delta}$ vanishes, the spacetime is said to be conformally flat.

1.6. Newman–Penrose Formalism

The NP formalism (Newman and Penrose, 1962, 1963) has been used successfully for both studying the asymptotic properties of a gravitational field (Penrose, 1960, 1963, 1964; Hawking, 1975) and obtaining solutions of the Einstein field equations (see, for example, Kinnersley, 1969; Talbot, 1969; Lind, 1974; Jagia and Griffiths, 1980; Kramer *et al.*, 1980; Singh and Grifffiths, 1990; Griffiths and Podolský, 2009; Stephani *et al.*, 2009). The language of the NP formalism has been generally accepted and widely used in the literature (see, for example, Pirani, 1964; Davies, 1976a, 1976b; Campbell and Wainwright, 1977; Frolov, 1979; Kramer *et al.*, 1980; Chandrasekhar, 1983; Stephani *et al.*, 2009). In these books, however, either the treatment is not consistent, for example, in Chandrasekhar's book, Eq. (287) is not consistent with Eqs. (288) and (289), or the definitions of the Riemann and Ricci tensors (Newman and Penrose, 1962; Frolov, 1979), or the signature of the metric (Kramer *et al.*, 1980; Stephani *et al.*, 2009) do not coincide with the ones that we use in this book. Thus, in this section we provide a complete and consistent form for the NP formalism. For a more detailed discussion, we refer the readers to the original Newman and Penrose papers (Newman and Penrose, 1962, 1963), Frolov (1979), Kramer *et al.* (1980), Wang (1991f) and Stephani *et al.* (2009).

The NP formalism is a tetrad formalism but with a special choice of the tetrad. NP chose the tetrad so that it consists of four null vectors, $l, n, m, \overline{m}$, denoted by

$$e^{\mu}_{(\alpha)} \equiv \{ l^{\mu}, n^{\mu}, m^{\mu}, \overline{m}^{\mu} \}, \tag{1.50}$$

where l^μ and n^μ are real, and m^μ and $\overline{m}^\mu$ are complex conjugates. The orthogonality properties of these vectors are

$$l^\mu l_\mu = n^\mu n_\mu = m^\mu m_\mu = \overline{m}^\mu \overline{m}_\mu = 0,$$

$$l^\mu m_\mu = l^\mu \overline{m}_\mu = n^\mu m_\mu = n^\mu \overline{m}_\mu = 0, \tag{1.51}$$

$$l^\mu n_\mu = -m^\mu \overline{m}_\mu = 1.$$

The tetrad indices can be raised and lowered by $\eta^{(\alpha)(\beta)}$ and $\eta_{(\alpha)(\beta)}$, respectively, which are given by

$$\left[\eta_{(\alpha)(\beta)}\right] = \left[\eta^{(\alpha)(\beta)}\right] = \begin{bmatrix} 0 & +1 & 0 & 0 \\ +1 & 0 & 0 & 0 \\ 0 & 0 & 0 & -1 \\ 0 & 0 & -1 & 0 \end{bmatrix}. \tag{1.52}$$

It is easy to show that

$$g_{\mu\nu} = \eta^{(\alpha)(\beta)} e_{\mu(\alpha)} e_{\nu(\beta)}$$

$$= l_\mu n_\nu + l_\nu n_\mu - m_\mu \overline{m}_\nu - m_\nu \overline{m}_\mu, \tag{1.53}$$

$$\eta_{(\alpha)(\beta)} = g^{\mu\nu} e_{\mu(\alpha)} e_{\nu(\beta)} = e^\mu_{(\alpha)} e_{\mu(\beta)}.$$

The complex spin coefficients $\gamma_{(\alpha)(\beta)(\gamma)}$ are defined by

$$\gamma_{(\alpha)(\beta)(\gamma)} \equiv e_{(\alpha)\mu;\nu} e^\mu_{(\beta)} e^\nu_{(\gamma)}, \tag{1.54}$$

or equivalently

$$e_{(\alpha)\mu;\nu} = \gamma_{(\alpha)(\beta)(\gamma)} e^{(\beta)}_\mu e^{(\gamma)}_\nu. \tag{1.55}$$

Since $\left[e^\mu_{(\beta)} e_{\mu(\gamma)}\right]_{;\nu} = 0$, it can be shown that the spin coefficients satisfy the relations

$$\gamma_{(\alpha)(\beta)(\gamma)} = -\gamma_{(\beta)(\alpha)(\gamma)}. \tag{1.56}$$

Note that the calculations of the spin coefficients does not require the calculations of the covariant derivatives. In fact, introducing the λ-symbols via the relations

$$\lambda_{(\alpha)(\beta)(\gamma)} \equiv \gamma_{(\alpha)(\beta)(\gamma)} - \gamma_{(\alpha)(\gamma)(\beta)}$$

$$= [e_{(\alpha)\mu;\nu} - e_{(\alpha)\nu;\mu}] e^\mu_{(\beta)} e^\nu_{(\gamma)}$$

$$= [e_{(\alpha)\mu,\nu} - e_{(\alpha)\nu,\mu}] e^\mu_{(\beta)} e^\nu_{(\gamma)}, \tag{1.57}$$

we find that the spin coefficients are given by

$$\gamma_{(\alpha)(\beta)(\gamma)} = \frac{1}{2}[\lambda_{(\alpha)(\beta)(\gamma)} + \lambda_{(\beta)(\gamma)(\alpha)} - \lambda_{(\gamma)(\alpha)(\beta)}]. \tag{1.58}$$

As is evident from Eq. (1.57), the calculations of $\lambda_{(\alpha)(\beta)(\gamma)}$ (consequently, the calculations of $\gamma_{(\alpha)(\beta)(\gamma)}$) require only the calculations of partial derivatives. On the other hand, from Eq. (1.57) we find that the λ-symbols have the properties,

$$\lambda_{(\alpha)(\beta)(\gamma)} = -\lambda_{(\alpha)(\gamma)(\beta)}. \tag{1.59}$$

The purpose of the NP formalism is to express the Einstein field equations in terms of the spin coefficients. To do so, we return to Eq. (1.12). By projecting the Ricci identity onto the tetrad frame, we have

$$\begin{aligned}
R_{(\alpha)(\beta)(\gamma)(\delta)} &= R_{\sigma\mu\nu\lambda} e^{\sigma}_{(\alpha)} e^{\mu}_{(\beta)} e^{\nu}_{(\gamma)} e^{\lambda}_{(\delta)} \\
&= [e_{(\alpha)\mu;\nu;\lambda} - e_{(\alpha)\mu;\lambda;\nu}] e^{\mu}_{(\beta)} e^{\nu}_{(\gamma)} e^{\lambda}_{(\delta)} \\
&= \gamma_{(\alpha)(\beta)(\gamma),(\delta)} - \gamma_{(\alpha)(\beta)(\delta),(\gamma)} \\
&\quad + \gamma_{(\alpha)(\beta)(\varepsilon)}[\gamma_{(\delta)}{}^{(\varepsilon)}{}_{(\gamma)} - \gamma_{(\gamma)}{}^{(\varepsilon)}{}_{(\delta)}] \\
&\quad + \gamma_{(\alpha)(\varepsilon)(\delta)}\gamma_{(\beta)}{}^{(\varepsilon)}{}_{(\gamma)} - \gamma_{(\alpha)(\varepsilon)(\gamma)}\gamma_{(\beta)}{}^{(\varepsilon)}{}_{(\delta)}, \tag{1.60}
\end{aligned}$$

where

$$\gamma_{(\alpha)}{}^{(\beta)}{}_{(\gamma)} \equiv \eta^{(\beta)(\delta)}\gamma_{(\alpha)(\delta)(\gamma)}, \quad \gamma_{(\alpha)(\beta)(\gamma),(\delta)} \equiv \gamma_{(\alpha)(\beta)(\gamma);\mu} e^{\mu}_{(\delta)}. \tag{1.61}$$

Since for any vector we have

$$\begin{aligned}
A_{(\alpha),(\beta)} &= A_{(\alpha);\nu} e^{\nu}_{(\beta)} = e^{\nu}_{(\beta)}(A_{\mu} e^{\mu}_{(\alpha)})_{;\nu} \\
&= e^{\mu}_{(\alpha)} A_{\mu;\nu} e^{\nu}_{(\beta)} + A_{(\varepsilon)}\gamma_{(\alpha)}{}^{(\varepsilon)}{}_{(\beta)}, \tag{1.62}
\end{aligned}$$

we can define the intrinsic derivative of $A_{(\alpha)}$ in the $e_{(\beta)}$-direction as

$$A_{(\alpha)|(\beta)} \equiv e^{\mu}(\alpha) A_{\mu;\nu} e^{\nu}_{(\beta)} = A_{(\alpha),(\beta)} - A_{(\varepsilon)}\gamma_{(\alpha)}{}^{(\varepsilon)}{}_{(\beta)}. \tag{1.63}$$

Similarly, we define $A^{(\alpha)}{}_{|(\beta)}$ as

$$A^{(\alpha)}{}_{|(\beta)} \equiv e^{(\alpha)}_{\mu} A^{\mu}{}_{;\nu} e^{\nu}_{(\beta)} = A^{(\alpha)}{}_{,(\beta)} + A^{(\varepsilon)}\gamma_{(\varepsilon)}{}^{(\alpha)}{}_{(\beta)}, \tag{1.64}$$

and for the more general case we have

$$
\begin{aligned}
R^{(\alpha)}{}_{(\beta)(\gamma)(\delta)|(\varepsilon)} &\equiv R^{\sigma}{}_{\mu\nu\lambda;\rho} e^{(\alpha)}_{\sigma} e^{\mu}_{(\beta)} e^{\nu}_{(\gamma)} \, e^{\,\lambda}_{\,(\delta)} \, e^{\,\rho}_{\,(\varepsilon)} \\
&= R^{(\alpha)}{}_{(\beta)(\gamma)(\delta),(\varepsilon)} + R^{(\zeta)}{}_{(\beta)(\gamma)(\delta)} \gamma^{(\alpha)}_{(\zeta)}{}_{(\varepsilon)} \\
&\quad - R^{(\alpha)}{}_{(\zeta)(\gamma)(\delta)} \gamma^{(\zeta)}_{(\beta)}{}_{(\varepsilon)} - R^{(\alpha)}{}_{(\beta)(\zeta)(\delta)} \gamma^{(\zeta)}_{(\gamma)}{}_{(\varepsilon)} \\
&\quad - R^{(\alpha)}{}_{(\beta)(\gamma)(\zeta)} \gamma^{(\zeta)}_{(\delta)}{}_{(\varepsilon)}.
\end{aligned}
\tag{1.65}
$$

Therefore, in terms of the intrinsic derivatives, the Bianchi identities (1.15) take the form

$$
R^{(\alpha)}{}_{(\beta)(\gamma)(\delta)|(\varepsilon)} + R^{(\alpha)}{}_{(\beta)(\varepsilon)(\gamma)|(\delta)} + R^{(\alpha)}{}_{(\beta)(\delta)(\varepsilon)|(\gamma)} = 0.
\tag{1.66}
$$

On the other hand, by projecting Eq. (1.49) onto the tetrad frame, we find that the relationship among the Riemann, Weyl and Ricci tensors goes over in the tetrad form without change, and thus is given by

$$
\begin{aligned}
R_{(\alpha)(\beta)(\gamma)(\delta)} &= C_{(\alpha)(\beta)(\gamma)(\delta)} + \frac{1}{2}\left[\eta_{(\alpha)(\gamma)} R_{(\beta)(\delta)}\right. \\
&\quad \left. + \eta_{(\beta)(\delta)} R_{(\alpha)(\gamma)} - \eta_{(\beta)(\gamma)} R_{(\alpha)(\delta)} - \eta_{(\alpha)(\delta)} R_{(\beta)(\gamma)}\right] \\
&\quad + \frac{1}{6}\left[\eta_{(\alpha)(\delta)}\eta_{(\beta)(\gamma)} - \eta_{(\alpha)(\gamma)}\eta_{(\beta)(\delta)}\right] R,
\end{aligned}
\tag{1.67}
$$

where

$$
R \equiv \eta^{(\alpha)(\beta)} R_{(\alpha)(\beta)} = 2\left[R_{(0)(1)} - R_{(2)(3)}\right].
\tag{1.68}
$$

Having written all of the formulas that we need in terms of the tetrad components, we are now ready to write down the NP equations. However, before we do so, following NP (Newman and Penrose, 1962), we introduce the following special notations that considerably simplify the expressions of the NP equations.

First of all, the spin coefficients are designated by

$$
\begin{aligned}
\kappa &\equiv \gamma_{(0)(2)(0)} = l_{\mu;\nu} m^{\mu} l^{\nu}, & \nu &\equiv -\gamma_{(1)(3)(1)} = -n_{\mu;\nu}\overline{m}^{\mu} n^{\nu}, \\
\rho &\equiv \gamma_{(0)(2)(3)} = l_{\mu;\nu} m^{\mu}\overline{m}^{\nu}, & \mu &\equiv -\gamma_{(1)(3)(2)} = -n_{\mu;\nu}\overline{m}^{\mu} m^{\nu}, \\
\sigma &\equiv \gamma_{(0)(2)(2)} = l_{\mu;\nu} m^{\mu} m^{\nu}, & \lambda &\equiv -\gamma_{(1)(3)(3)} = -n_{\mu;\nu}\overline{m}^{\mu}\overline{m}^{\nu}, \\
\tau &\equiv \gamma_{(0)(2)(1)} = l_{\mu;\nu} m^{\mu} n^{\nu}, & \pi &\equiv -\gamma_{(1)(3)(0)} = -n_{\mu;\nu}\overline{m}^{\mu} l^{\nu},
\end{aligned}
$$

$$\varepsilon \equiv \frac{1}{2}[\gamma_{(0)(1)(0)} - \gamma_{(2)(3)(0)}] = \frac{1}{2}[l_{\mu;\nu}n^\mu l^\nu - m_{\mu;\nu}\overline{m}^\mu l^\nu],$$

$$\alpha \equiv \frac{1}{2}[\gamma_{(0)(1)(3)} - \gamma_{(2)(3)(3)}] = \frac{1}{2}[l_{\mu;\nu}n^\mu \overline{m}^\nu - m_{\mu;\nu}\overline{m}^\mu \overline{m}^\nu],$$

$$\beta \equiv \frac{1}{2}[\gamma_{(0)(1)(2)} - \gamma_{(2)(3)(2)}] = \frac{1}{2}[l_{\mu;\nu}n^\mu m^\nu - m_{\mu;\nu}\overline{m}^\mu m^\nu],$$

$$\gamma \equiv \frac{1}{2}[\gamma_{(0)(1)(1)} - \gamma_{(2)(3)(1)}] = \frac{1}{2}[l_{\mu;\nu}n^\mu n^\nu - m_{\mu;\nu}\overline{m}^\mu n^\nu], \qquad (1.69)$$

and all other spin coefficients can be obtained from them by using the symmetry given by Eq. (1.56) and the fact that the complex conjugate of any quantity can be obtained by replacing the index 2, wherever it appears, by the index 3, and vice versa.

As mentioned previously, the Weyl tensor has 10 independent components at each point of spacetime. In the NP formalism, these components are specified by five complex "scalars" as follows:

$$
\begin{aligned}
C_{\mu\nu\lambda\delta} \equiv\ & -4\left(\Psi_2 + \overline{\Psi}_2\right)[l_{[\mu}n_{\nu]}l_{[\lambda}n_{\delta]} + m_{[\mu}\overline{m}_{\nu]}m_{[\lambda}\overline{m}_{\delta]}] \\
& + 4\left(\Psi_2 - \overline{\Psi}_2\right)[1_{[\mu}n_{\nu]}m_{[\lambda}\overline{m}_{\delta]} + m_{[\mu}\overline{m}_{\nu]}l_{[\lambda}n_{\delta]}] \\
& - 4\{\Psi_0 n_{[\mu}\overline{m}_{\nu]}n_{[\lambda}\overline{m}_{\delta]} + \Psi_1[l_{[\mu}n_{\nu]}n_{[\lambda}\overline{m}_{\delta]} + n_{[\mu}\overline{m}_{\nu]}l_{[\lambda}n_{\delta]} \\
& + n_{[\mu}\overline{m}_{\nu]}\overline{m}_{[\lambda}m_{\delta]} + \overline{m}_{[\mu}m_{\nu]}n_{[\lambda}\overline{m}_{\delta]}] - \Psi_2[l_{[\mu}m_{\nu]}n_{[\lambda}\overline{m}_{\delta]} \\
& + n_{[\mu}\overline{m}_{\nu]}l_{[\lambda}m_{\delta]}] - \Psi_3[l_{[\mu}n_{\nu]}l_{[\lambda}m_{\delta]} + l_{[\mu}m_{\nu]}l_{[\lambda}n_{\delta]} \\
& - l_{[\mu}m_{\nu]}m_{[\lambda}\overline{m}_{\delta]} - m_{[\mu}\overline{m}_{\nu]}l_{[\lambda}m_{\delta]}] + \Psi_4 l_{[\mu}m_{\nu]}l_{[\lambda}m_{\delta]} \\
& + \text{Complex conjugates}\},
\end{aligned}
\qquad (1.70)
$$

where $\Psi_0, \ldots, \Psi_4$ are called the Weyl scalars, and a bar over a letter denotes the complex conjugate, as mentioned previously.

The various terms in Eq. (1.70) have the following physical interpretations (Szekeres, 1965). The Ψ_0 and Ψ_1 terms represent, respectively, the transverse and longitudinal wave components in the n^μ direction, the Ψ_2 term a "Coulomb" component, and the Ψ_3 and Ψ_4 terms the longitudinal and transverse wave components in the l^μ direction. By contracting Eq. (1.70) with appropriate combinations of the null vectors, $\{l, n, m, \overline{m}\}$, we find

$$\Psi_0 \equiv -C_{(0)(2)(0)(2)} = -C_{\mu\nu\lambda\delta}l^\mu m^\nu l^\lambda m^\delta,$$

$$\Psi_1 \equiv -C_{(0)(1)(0)(2)} = -C_{\mu\nu\lambda\delta}1^\mu n^\nu l^\lambda m^\delta,$$

$$\Psi_2 \equiv -\frac{1}{2}[C_{(0)(1)(0)(1)} - C_{(0)(1)(2)(3)}]$$

$$= -\frac{1}{2}C_{\mu\nu\lambda\delta}[l^\mu n^\nu l^\lambda n^\delta - l^\mu n^\nu m^\lambda \overline{m}^\delta],$$

$$\Psi_3 \equiv C_{(0)(1)(1)(3)} = C_{\mu\nu\lambda\delta}l^\mu n^\nu n^\lambda \overline{m}^\delta$$

$$= -C_{\mu\nu\lambda\delta}n^\mu l^\nu n^\lambda \overline{m}^\delta,$$

$$\Psi_4 \equiv -C_{(1)(3)(1)(3)} = -C_{\mu\nu\lambda\delta}n^\mu \overline{m}^\nu n^\lambda \overline{m}^\delta. \tag{1.71}$$

Similarly, the tetrad components of the traceless Ricci tensor, $S_{(\alpha)(\beta)}$, can be written by using the following notation:

$$\Phi_{00} \equiv \frac{1}{2}S_{(0)(0)} = \frac{1}{2}S_{\mu\nu}l^\mu l^\nu = \overline{\Phi}_{00} = \frac{1}{2}R_{(0)(0)},$$

$$\Phi_{01} \equiv \frac{1}{2}S_{(0)(2)} = \frac{1}{2}S_{\mu\nu}l^\mu m^\nu = \overline{\Phi}_{10} = \frac{1}{2}R_{(0)(2)},$$

$$\Phi_{02} \equiv \frac{1}{2}S_{(2)(2)} = \frac{1}{2}S_{\mu\nu}m^\mu m^\nu = \overline{\Phi}_{20} = \frac{1}{2}R_{(2)(2)},$$

$$\Phi_{11} \equiv \frac{1}{4}[S_{(0)(1)} + S_{(2)(3)}] = \frac{1}{4}S_{\mu\nu}\left(l^\mu n^\nu + m^\mu \overline{m}^\nu\right) \tag{1.72}$$

$$= \overline{\Phi}_{11} = \frac{1}{4}[R_{(0)(1)} + R_{(2)(3)}],$$

$$\Phi_{12} \equiv \frac{1}{2}S_{(1)(2)} = \frac{1}{2}S_{\mu\nu}n^\mu m^\nu = \overline{\Phi}_{21} = \frac{1}{2}R_{(1)(2)},$$

$$\Phi_{22} \equiv \frac{1}{2}S_{(1)(1)} = \frac{1}{2}S_{\mu\nu}n^\mu n^\nu = \overline{\Phi}_{22} = \frac{1}{2}R_{(1)(1)},$$

and the trace of the Ricci tensor $R_{(\alpha)(\beta)}$ is defined by

$$\Lambda \equiv -\frac{1}{24}R = -\frac{1}{12}[R_{(0)(1)} - R_{(2)(3)}]. \tag{1.73}$$

We hope that there will be no confusion between the cosmological constant Λ used in Eq. (1.20) and the one used in Eq. (1.73), as in this book we shall mainly consider the cases in which the cosmological constant vanishes.[3]

---

[3]An interesting mechanism was recently proposed to produce a cosmological constant by the collision of two plane gravitational and/or electromagnetic waves (Barrabés and Hogan, 2014, 2015; Halisoy, Mazharimousavi and Gurtug, 2014).

In terms of Φ_{ij} and Λ, the energy–momentum tensor takes the form,

$$
\begin{aligned}
T_{\mu\nu} = 2\{ & \Phi_{22} l_\mu l_\nu + \Phi_{00} n_\mu n_\nu + (\Phi_{11} + 3\Lambda)(l_\mu n_\nu + l_\nu n_\mu) \\
& + (\Phi_{11} - 3\Lambda)(m_\mu \overline{m}_\nu + m_\nu \overline{m}_\mu) \\
& - \Phi_{01}(n_\mu \overline{m}_\nu + n_\nu \overline{m}_\mu) - \overline{\Phi}_{01}(n_\mu m_\nu + n_\nu m_\mu) \\
& - \Phi_{12}(l_\mu \overline{m}_\nu + l_\nu \overline{m}_\mu) - \overline{\Phi}_{12}(l_\mu m_\nu + l_\nu m_\mu) \\
& + \Phi_{02} \overline{m}_\mu \overline{m}_\nu + \overline{\Phi}_{02} m_\mu m_\nu \}.
\end{aligned}
\tag{1.74}
$$

On the other hand, the combination of Eq. (1.67) and Eqs. (1.70)–(1.73) gives

$$
\begin{aligned}
R_{(0)(1)(0)(1)} &= C_{(0)(1)(0)(1)} - R_{(0)(1)} + \frac{1}{6}R \\
&= -\left(\Psi_2 + \overline{\Psi}_2\right) - 2\Phi_{11} + 2\Lambda, \\[2mm]
R_{(0)(1)(0)(2)} &= C_{(0)(1)(0)(2)} - \frac{1}{2}R_{(0)(2)} = -\Psi_1 - \Phi_{01}, \\[2mm]
R_{(0)(1)(1)(2)} &= C_{(0)(1)(1)(2)} + \frac{1}{2}R_{(1)(2)} = \overline{\Psi}_3 + \Phi_{12}, \\[2mm]
R_{(0)(1)(2)(3)} &= C_{(0)(1)(2)(3)} = \Psi_2 - \overline{\Psi}_2, \\[2mm]
R_{(0)(2)(0)(2)} &= C_{(0)(2)(0)(2)} = -\Psi_0, \\[2mm]
R_{(0)(2)(0)(3)} &= -\frac{1}{2}R_{(0)(0)} = -\Phi_{00}, \\[2mm]
R_{(0)(2)(1)(2)} &= \frac{1}{2}R_{(2)(2)} = \Phi_{02}, \\[2mm]
R_{(0)(2)(1)(3)} &= C_{(0)(2)(1)(3)} - \frac{1}{12}R = \Psi_2 + 2\Lambda, \\[2mm]
R_{(0)(2)(2)(3)} &= C_{(0)(2)(2)(3)} - \frac{1}{2}R_{(0)(2)} = \Psi_1 - \Phi_{01}, \\[2mm]
R_{(1)(2)(1)(2)} &= C_{(1)(2)(1)(2)} = -\overline{\Psi}_4, \\[2mm]
R_{(1)(2)(1)(3)} &= -\frac{1}{2}R_{(1)(1)} = -\Phi_{22}, \\[2mm]
R_{(1)(2)(2)(3)} &= C_{(1)(2)(2)(3)} - \frac{1}{2}R_{(1)(2)} = \Psi_3 - \Phi_{12}, \\[2mm]
R_{(2)(3)(2)(3)} &= C_{(2)(3)(2)(3)} + R_{(2)(3)} + \frac{1}{6}R \\
&= -\left(\Psi_2 + \overline{\Psi}_2\right) + 2\Phi_{11} + 2\Lambda.
\end{aligned}
\tag{1.75}
$$

We also define the operators,

$$D \equiv l^\mu \partial_\mu, \quad \Delta \equiv n^\mu \partial_\mu, \quad \delta \equiv m^\mu \partial_\mu, \quad \overline{\delta} \equiv \overline{m}^\mu \partial_\mu. \tag{1.76}$$

Substituting Eqs. (1.69), (1.75) and (1.76) into Eq. (1.65), we find

$$D\rho - \overline{\delta}\kappa = (\rho^2 + \sigma\overline{\sigma}) + \rho(\varepsilon + \overline{\varepsilon}) - \kappa(3\alpha + \overline{\beta} - \pi) - \overline{\kappa}\tau + \Phi_{00}, \tag{1.77a}$$

$$D\sigma - \delta\kappa = \sigma(\rho + \overline{\rho} + 3\varepsilon - \overline{\varepsilon}) - \kappa(\tau - \overline{\pi} + \overline{\alpha} + 3\beta) + \Psi_0, \tag{1.77b}$$

$$D\tau - \Delta\kappa = \rho(\tau + \overline{\pi}) + \sigma(\overline{\tau} + \pi) + \tau(\varepsilon - \overline{\varepsilon})$$
$$- \kappa(3\gamma + \overline{\gamma}) + \Psi_1 + \Phi_{01}, \tag{1.77c}$$

$$D\alpha - \overline{\delta}\varepsilon = \alpha(\rho + \overline{\varepsilon} - 2\varepsilon) + \pi(\varepsilon + \rho) + \beta\overline{\sigma} - \overline{\beta}\varepsilon$$
$$- \kappa\lambda - \overline{\kappa}\gamma + \Phi_{10}, \tag{1.77d}$$

$$D\beta - \delta\varepsilon = \sigma(\alpha + \pi) + \beta(\overline{\rho} - \overline{\varepsilon}) - \kappa(\mu + \gamma) - \varepsilon(\overline{\alpha} - \overline{\pi}) + \Psi_1, \tag{1.77e}$$

$$D\gamma - \Delta\varepsilon = \alpha(\tau + \overline{\pi}) + \beta(\overline{\tau} + \pi) - \gamma(\varepsilon + \overline{\varepsilon}) - \varepsilon(\gamma + \overline{\gamma})$$
$$+ \tau\pi - \nu\kappa + \Psi_2 + \Phi_{11} - \Lambda, \tag{1.77f}$$

$$D\lambda - \overline{\delta}\pi = \pi(\alpha + \pi - \overline{\beta}) - \lambda(3\varepsilon - \overline{\varepsilon}) + \rho\lambda + \overline{\sigma}\mu - \nu\overline{\kappa} + \Phi_{20}, \tag{1.77g}$$

$$D\mu - \delta\pi = \pi(\overline{\pi} - \overline{\alpha} + \beta) - \mu(\varepsilon + \overline{\varepsilon}) + \overline{\rho}\mu$$
$$+ \sigma\lambda - \nu\kappa + \Psi_2 + 2\Lambda, \tag{1.77h}$$

$$D\nu - \Delta\pi = \mu(\pi + \overline{\tau}) + \lambda(\overline{\pi} + \tau) + \pi(\gamma - \overline{\gamma})$$
$$- \nu(3\varepsilon + \overline{\varepsilon}) + \Psi_3 + \Phi_{21}, \tag{1.77i}$$

$$\Delta\lambda - \overline{\delta}\nu = -\lambda(\mu + \overline{\mu} + 3\gamma - \overline{\gamma}) + \nu(3\alpha + \overline{\beta} + \pi - \overline{\tau}) - \Psi_4, \tag{1.77j}$$

$$\delta\rho - \overline{\delta}\sigma = \rho(\overline{\alpha} + \beta) - \sigma(3\alpha - \overline{\beta}) + \tau(\rho - \overline{\rho})$$
$$+ \kappa(\mu - \overline{\mu}) - \Psi_1 + \Phi_{01}, \tag{1.77k}$$

$$\delta\alpha - \overline{\delta}\beta = \gamma(\rho - \overline{\rho}) + \varepsilon(\mu - \overline{\mu}) + \mu\rho - \lambda\sigma + \alpha\overline{\alpha} + \beta\overline{\beta}$$
$$- 2\alpha\beta - \Psi_2 + \Phi_{11} + \Lambda, \tag{1.77l}$$

$$\delta\lambda - \overline{\delta}\mu = \nu(\rho - \overline{\rho}) + \pi(\mu - \overline{\mu}) + \mu(\alpha + \overline{\beta})$$
$$+ \lambda(\overline{\alpha} - 3\beta) - \Psi_3 + \Phi_{21}, \tag{1.77m}$$

$$\delta\nu - \Delta\mu = \nu(\tau - 3\beta - \overline{\alpha}) + \mu(\gamma + \overline{\gamma}) - \overline{\nu}\pi + \mu^2 + \lambda\overline{\lambda} + \Phi_{22}, \tag{1.77n}$$

$$\delta\gamma - \Delta\beta = \gamma(\tau - \overline{\alpha} - \beta) - \beta(\gamma - \overline{\gamma} - \mu)$$
$$+ \mu\tau - \sigma\nu - \varepsilon\overline{\nu} + \alpha\overline{\lambda} + \Phi_{12}, \tag{1.77o}$$

$$\delta\tau - \Delta\sigma = \tau(\tau + \beta - \overline{\alpha}) - \sigma(3\gamma - \overline{\gamma}) + \mu\sigma + \overline{\lambda}\rho - \kappa\overline{\nu} + \Phi_{02}, \quad (1.77\text{p})$$

$$\Delta\rho - \overline{\delta}\tau = \tau(\overline{\beta} - \alpha - \overline{\tau}) + \rho(\gamma + \overline{\gamma}) - \overline{\mu}\rho$$

$$- \lambda\sigma + \nu\kappa - \Psi_2 - 2\Lambda, \quad (1.77\text{q})$$

$$\Delta\alpha - \overline{\delta}\gamma = \nu(\rho + \varepsilon) - \lambda(\tau + \beta) + \alpha(\overline{\gamma} - \overline{\mu}) + \gamma(\overline{\beta} - \overline{\tau}) - \Psi_3. \quad (1.77\text{r})$$

Combining Eqs. (1.65)–(1.69) and Eqs. (1.71)–(1.76), on the other hand, we find that the Bianchi identities read

$$\overline{\delta}\Psi_0 - D\Psi_1 + D\Phi_{01} - \delta\Phi_{00}$$

$$= (4\alpha - \pi)\Psi_0 - 2(2\rho + \varepsilon)\Psi_1 + 3\kappa\Psi_2 + (\overline{\pi} - 2\overline{\alpha} - 2\beta)\Phi_{00}$$

$$+ 2(\varepsilon + \overline{\rho})\Phi_{01} + 2\sigma\Phi_{10} - 2\kappa\Phi_{11} - \overline{\kappa}\Phi_{02}, \quad (1.78\text{a})$$

$$\Delta\Psi_0 - \delta\Psi_1 + D\Phi_{02} - \delta\Phi_{01}$$

$$= (4\gamma - \mu)\Psi_0 - 2(2\tau + \beta)\Psi_1 + 3\sigma\Psi_2 + (2\varepsilon - 2\overline{\varepsilon} + \overline{\rho})\Phi_{02}$$

$$+ 2(\overline{\pi} - \beta)\Phi_{01} + 2\sigma\Phi_{11} - 2\kappa\Phi_{12} - \overline{\lambda}\Phi_{00}, \quad (1.78\text{b})$$

$$3\left(\overline{\delta}\Psi_1 - D\Psi_2\right) + 2\left(D\Phi_{11} - \delta\Phi_{10}\right) + \overline{\delta}\Phi_{01} - \Delta\Phi_{00}$$

$$= 3\lambda\Psi_0 - 9\rho\Psi_2 + 6(\alpha - \pi)\Psi_1 + 6\kappa\Psi_3 + (\overline{\mu} - 2\mu - 2\gamma - 2\overline{\gamma})\Phi_{00}$$

$$+ 2(\alpha + \pi + \overline{\tau})\Phi_{01} + 2(\tau - 2\overline{\alpha} + \overline{\pi})\Phi_{10} + 2(2\overline{\rho} - \rho)\Phi_{11}$$

$$+ 2\sigma\Phi_{20} - 2\overline{\kappa}\Phi_{12} - 2\kappa\Phi_{21} - \overline{\sigma}\Phi_{02}, \quad (1.78\text{c})$$

$$3\left(\Delta\Psi_1 - \delta\Psi_2\right) + 2\left(D\Phi_{12} - \delta\Phi_{11}\right) + \overline{\delta}\Phi_{02} - \Delta\Phi_{01}$$

$$= 3\nu\Psi_0 - 9\tau\Psi_2 + 6(\gamma - \mu)\Psi_1 + 6\sigma\Psi_3 + (2\alpha + 2\pi + \overline{\tau} - 2\overline{\beta})\Phi_{02}$$

$$+ 2(\overline{\rho} - \rho - 2\overline{\varepsilon})\Phi_{12} + 2(\overline{\mu} - \mu - \gamma)\Phi_{01} + 2(\tau + 2\overline{\pi})\Phi_{11}$$

$$+ 2\sigma\Phi_{21} - 2\overline{\lambda}\Phi_{10} - 2\kappa\Phi_{22} - \overline{\nu}\Phi_{00}, \quad (1.78\text{d})$$

$$3\left(\overline{\delta}\Psi_2 - D\Psi_3\right) + 2\left(\overline{\delta}\Phi_{11} - \Delta\Phi_{10}\right) + D\Phi_{21} - \delta\Phi_{20}$$

$$= 3\kappa\Psi_4 - 9\pi\Psi_2 + 6(\varepsilon - \rho)\Psi_3 + 6\lambda\Psi_1 + (2\beta + 2\tau + \overline{\pi} - 2\overline{\alpha})\Phi_{20}$$

$$+ 2(\overline{\mu} - \mu - 2\overline{\gamma})\Phi_{10} + 2(\overline{\rho} - \rho - \varepsilon)\Phi_{21} + 2(\pi + 2\overline{\tau})\Phi_{11}$$

$$+ 2\lambda\Phi_{01} - 2\overline{\sigma}\Phi_{12} - 2\nu\Phi_{00} - \overline{\kappa}\Phi_{22}, \quad (1.78\text{e})$$

$$3\left(\Delta\Psi_2 - \delta\Psi_3\right) + 2\left(\overline{\delta}\Phi_{12} - \Delta\Phi_{11}\right) + D\Phi_{22} - \delta\Phi_{21}$$

$$= 3\sigma\Psi_4 - 9\mu\Psi_2 + 6(\beta - \tau)\Psi_3 + 6\nu\Psi_1 + (\overline{\rho} - 2\varepsilon - 2\overline{\varepsilon} - 2\rho)\Phi_{22}$$

$$+ 2(\pi + \overline{\tau} - 2\overline{\beta})\Phi_{12} + 2(\beta + \tau + \overline{\pi})\Phi_{21} + 2(2\overline{\mu} - \mu)\Phi_{11}$$

$$+ 2\lambda\Phi_{02} - 2\nu\Phi_{01} - 2\overline{\nu}\Phi_{10} - \overline{\lambda}\Phi_{20}, \tag{1.78f}$$

$$\overline{\delta}\Psi_3 - D\Psi_4 + \overline{\delta}\Phi_{21} - \Delta\Phi_{20}$$

$$= (4\varepsilon - \rho)\Psi_4 - 2(2\pi + \alpha)\Psi_3 + 3\lambda\Psi_2 + (2\gamma - 2\overline{\gamma} + \overline{\mu})\Phi_{20}$$

$$+ 2(\overline{\tau} - \alpha)\Phi_{21} + 2\lambda\Phi_{11} - 2\nu\Phi_{10} - \overline{\sigma}\Phi_{22}, \tag{1.78g}$$

$$\Delta\Psi_3 - \delta\Psi_4 - \Delta\Phi_{21} + \overline{\delta}\Phi_{22}$$

$$= (4\beta - \tau)\Psi_4 - 2(2\mu + \gamma)\Psi_3 + 3\nu\Psi_2 + (\overline{\tau} - 2\alpha - 2\overline{\beta})\Phi_{22}$$

$$+ 2(\gamma + \overline{\mu})\Phi_{21} + 2\lambda\Phi_{12} - 2\nu\Phi_{11} - \overline{\nu}\Phi_{20}, \tag{1.78h}$$

$$D\Phi_{11} - \delta\Phi_{10} - \overline{\delta}\Phi_{01} + \Delta\Phi_{00} + 3D\Lambda$$

$$= (2\gamma + 2\overline{\gamma} - \mu - \overline{\mu})\Phi_{00} + (\pi - 2\alpha - 2\overline{\tau})\Phi_{01} + (\overline{\pi} - 2\overline{\alpha} - 2\tau)\Phi_{10}$$

$$+ 2(\rho + \overline{\rho})\Phi_{11} + \overline{\sigma}\Phi_{02} + \sigma\Phi_{20} - \overline{\kappa}\Phi_{12} - \kappa\Phi_{21}, \tag{1.78i}$$

$$D\Phi_{12} - \delta\Phi_{11} - \overline{\delta}\Phi_{02} + \Delta\Phi_{01} + 3\delta\Lambda$$

$$= (2\overline{\beta} - 2\alpha + \pi - \overline{\tau})\Phi_{02} + (2\rho + \overline{\rho} - 2\overline{\varepsilon})\Phi_{12} + (2\gamma - 2\overline{\mu} - \mu)\Phi_{01}$$

$$+ 2(\overline{\pi} - \tau)\Phi_{11} + \overline{\nu}\Phi_{00} + \sigma\Phi_{21} - \overline{\lambda}\Phi_{10} - \kappa\Phi_{22}, \tag{1.78j}$$

$$D\Phi_{22} - \delta\Phi_{21} - \overline{\delta}\Phi_{12} + \Delta\Phi_{11} + 3\Delta\Lambda$$

$$= (\rho + \overline{\rho} - 2\varepsilon - 2\overline{\varepsilon})\Phi_{22} + (2\pi - \overline{\tau} + 2\overline{\beta})\Phi_{12} + (2\beta - \tau + 2\overline{\pi})\Phi_{21}$$

$$- 2(\mu + \overline{\mu})\Phi_{11} + \nu\Phi_{01} + \overline{\nu}\Phi_{10} - \lambda\Phi_{02} - \overline{\lambda}\Phi_{20}. \tag{1.78k}$$

In terms of Φ_{ij}, Λ and Ψ_i, the Kretschmann scalar is given by

$$I \equiv R_{\alpha\beta\gamma\delta}R^{\alpha\beta\gamma\delta}$$

$$= 8\{3(\Psi_2{}^2 + \overline{\Psi}_2{}^2) - 4(\Psi_1\Psi_3 + \overline{\Psi}_1\overline{\Psi}_3) + (\Psi_0\Psi_4 + \overline{\Psi}_0\overline{\Psi}_4)$$

$$- 4(\Phi_{01}\overline{\Phi}_{12} + \overline{\Phi}_{01}\Phi_{12}) + 4\Phi_{11}{}^2 + 2\Phi_{02}\overline{\Phi}_{02}$$

$$+ 2\Phi_{00}\Phi_{22} + 12\Lambda^2\}. \tag{1.79}$$

1.7. Optical Scalars

The geometric meaning of the spin coefficients introduced in Section 1.6 is manifested from the study of the null congruences formed by l^μ and n^μ, respectively. Combining Eqs. (1.55) and (1.69), we find that

$$l_{\mu;\nu} = \gamma_{(0)(\beta)(\gamma)}e_\mu^{(\beta)}e_\nu^{(\gamma)} = (\gamma + \overline{\gamma})l_\mu l_\nu + (\varepsilon + \overline{\varepsilon})l_\mu n_\nu - (\alpha + \overline{\beta})l_\mu m_\nu$$
$$- (\beta + \overline{\alpha})l_\mu \overline{m}_\nu - \overline{\tau}m_\mu l_\nu - \overline{\kappa}m_\mu n_\nu + \overline{\sigma}m_\mu m_\nu + \overline{\rho}m_\mu \overline{m}_\nu$$
$$- \tau\overline{m}_\mu l_\nu - \kappa\overline{m}_\mu n_\nu + \sigma\overline{m}_\mu \overline{m}_\nu + \rho\overline{m}_\mu m_\nu, \tag{1.80}$$

and

$$n_{\mu;\nu} = \gamma_{(1)(\beta)(\gamma)}\,e_\mu^{(\beta)}e_\nu^{(\gamma)} = -(\gamma + \overline{\gamma})n_\mu l_\nu - (\varepsilon + \overline{\varepsilon})n_\mu n_\nu + (\alpha + \overline{\beta})n_\mu m_\nu$$
$$+ (\beta + \overline{\alpha})n_\mu \overline{m}_\nu + \nu m_\mu l_\nu + \pi m_\mu n_\nu - \lambda m_\mu m_\nu - \mu m_\mu \overline{m}_\nu + \overline{\nu}\,\overline{m}_\mu l_\nu$$
$$+ \overline{\pi}\,\overline{m}_\mu n_\nu - \overline{\lambda}\,\overline{m}_\mu \overline{m}_\nu - \overline{\mu}\,\overline{m}_\mu m_\nu. \tag{1.81}$$

Hence, we have

$$l_{\mu;\nu}l^\nu = (\varepsilon + \overline{\varepsilon})l_\mu - \overline{\kappa}m_\mu - \kappa\overline{m}_\mu,$$
$$n_{\mu;\nu}n^\nu = -(\gamma + \overline{\gamma})n_\mu + \nu m_\mu + \overline{\nu}\,\overline{m}_\mu. \tag{1.82}$$

Thus, if $\kappa = 0$, then l^μ defines a null geodesic congruence. If, in addition, $\mathrm{Re}(\varepsilon) = 0$, then l^μ is the tangent vector corresponding to an affine parameterization. The same holds for the null vector n^μ if κ and ε are replaced by $-\overline{\nu}$ and $-\overline{\gamma}$, respectively. Because of the symmetry between l^μ and n^μ, it is sufficient to consider only the null congruence formed by l^μ. And for the sake of convenience, we assume that this null congruence is an affinely parameterized geodesic congruence, i.e.

$$\kappa = 0 = \mathrm{Re}(\varepsilon). \tag{1.83}$$

Let S_O and S_P be infinitesimal two-dimensional surfaces spanned by **m** and $\overline{\mathbf{m}}$ and orthogonal to the null geodesic C at neighboring points O and P of C, and let S be an infinitesimal circle with center O, lying in S_O (see Fig. 1.2(a)). Suppose that the null geodesic congruence meets S_O in the circle S, then let us observe the image of this null congruence on S_P.

We first consider the deviation vector between the null geodesic C and any other, say, C', which meets the circle S at O' (Fig. 1.2(a)). Without loss of generality, we assume that η^μ lies on S_O. Then, from Eqs. (1.27) and (1.80) we find that the relative velocity of the two null geodesics C and C' is given by

$$\frac{D\eta^\mu}{D\lambda} = \eta^\mu_{\ ;\nu}l^\nu = l^\mu_{\ ;\nu}\eta^\nu$$
$$= (m_\nu\eta^\nu)\left[-(\alpha + \overline{\beta})l^\mu + \rho\overline{m}^\mu + \overline{\sigma}m^\mu\right]$$
$$+ (\overline{m}_\nu\eta^\nu)\left[-(\overline{\alpha} + \beta)l^\mu + \overline{\rho}m^\mu + \sigma\overline{m}^\mu\right]. \tag{1.84}$$

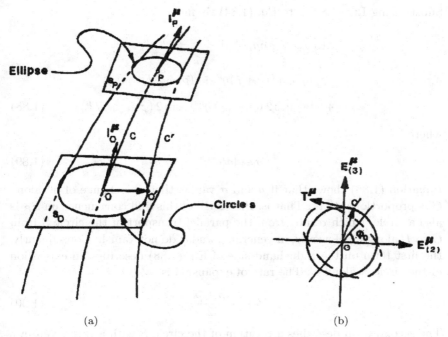

Fig. 1.2.   A null congruence meets S_o in the circle S. The image of the circle S on S_P is an ellipse.

By using **m** and $\overline{\mathbf{m}}$, we construct two space-like unit vectors via the relations

$$E_{(2)}{}^{\mu} = \frac{m^{\mu} + \overline{m}^{\mu}}{\sqrt{2}}, \qquad E_{(3)}{}^{\mu} = \frac{m^{\mu} - \overline{m}^{\mu}}{i\sqrt{2}}. \tag{1.85}$$

Then, it is easy to show that

$$\begin{aligned} E_{(2)}{}^{\mu} &= -\sin\varphi\, T^{\mu} + \cos\varphi\, \eta^{\mu}, \\ E_{(3)}{}^{\mu} &= \cos\varphi\, T^{\mu} + \sin\varphi\, \eta^{\mu}, \end{aligned} \tag{1.86}$$

where φ is the angle formed by η^{μ} and $E_{(2)}{}^{\mu}$ [Fig. 1.2(b)], and T^{μ} is the unit vector tangent to the circle S at O'. Thus we have

$$\begin{aligned} m^{\mu} &= \frac{E_{(2)}{}^{\mu} + iE_{(3)}{}^{\mu}}{\sqrt{2}} = \frac{e^{i\varphi}}{\sqrt{2}}\left(\eta^{\mu} + iT^{\mu}\right), \\ m^{\mu} &= \frac{E_{(2)}{}^{\mu} - iE_{(3)}{}^{\mu}}{\sqrt{2}} = \frac{e^{-i\varphi}}{\sqrt{2}}\left(\eta^{\mu} - iT^{\mu}\right). \end{aligned} \tag{1.87}$$

Substituting Eq. (1.87) into Eq. (1.84) we find

$$
\frac{D\eta^\mu}{D\lambda} + 2\operatorname{Re}[(\alpha + \overline{\beta})m_\nu \eta^\nu]l^\mu
$$

$$
= -\frac{1}{2}|\eta^\nu \eta_\nu|\{(\rho + \overline{\rho})\eta^\mu + i(\overline{\rho} - \rho)T^\mu
$$

$$
+ 2|\sigma|\left[\cos 2\left(\varphi - \varphi_0\right)\eta^\mu - \sin 2\left(\varphi - \varphi_0\right)T^\mu\right]\}, \qquad (1.88)
$$

where

$$
\sigma \equiv |\sigma|e^{i2\varphi_0}. \qquad (1.89)
$$

Equation (1.88) shows that if ρ and σ vanish then the change of η^μ along C is proportional to l^μ. That is, the image of this null congruence on S_P is also a circle, which results from the parallel transport of the circle S from O to P along C. However, in general, ρ and σ do not vanish. Consequently, the first term on the right-hand side of Eq. (1.88) describes an expansion of the circle S along C. The rate of expansion is

$$
\theta \equiv -\frac{\rho + \overline{\rho}}{2}. \qquad (1.90)
$$

The second term describes a rotation of the circle S with a rate given by

$$
\omega \equiv \frac{i(\rho - \overline{\rho})}{2}. \qquad (1.91)
$$

And the last term on the right-hand side of Eq. (1.88) depends explicitly on the angle φ. It is easy to see that because of this dependence the circle S goes over into an ellipse, and the minor axis of which forms an angle φ_0 with respect to $E_{(2)}{}^\mu$ [Fig. 1.2(b)].

For the null geodesic congruence formed by n^μ, the coefficients $-\overline{\nu}$, $-\overline{\gamma}$, $-\overline{\mu}$ and $-\overline{\lambda}$ correspond to κ, ε, ρ and σ, respectively. The quantities θ, ω and σ, as defined above, were first introduced by Sachs (1961, 1962, 1964), and are called the optical scalars.

Finally, we note that the change of the shape of a null congruence is due not only to the spacetime curvature, but also to the inertial field. The latter follows from the fact that even in a flat spacetime the optical scalars may not vanish because of the choice of the coordinate system. In order to consider the effects only due to the spacetime curvature, we must consider the variation of the optical scalars along geodesics (see Section 1.4). That is, we must consider the acceleration between geodesics, instead of considering their relative velocity.

1.8. Matter Fields

In this book, in addition to the considerations of exact solutions of the Einstein vacuum equations, we shall also consider solutions of the Einstein field equations for the following physically relevant energy–stress tensors.

(α) *A massless scalar field*: The energy–stress tensor for a massless scalar field ϕ takes the form

$$T_{\mu\nu} = \phi_{;\mu}\phi_{;\nu} - \frac{1}{2}g_{\mu\nu}\phi_{;\lambda}\phi^{;\lambda}, \tag{1.92}$$

where ϕ satisfies the massless Klein–Gordon equation

$$\phi_{;\mu;\nu}g^{\mu\nu} = 0. \tag{1.93}$$

(β) *A pure radiation field*: The energy–stress tensor in this case is given by

$$T_{\mu\nu} = \varepsilon\kappa_{\mu}\kappa_{\nu}, \qquad \kappa^{\nu}\kappa_{\nu} = 0, \tag{1.94}$$

where ε is non-negative.

Note that the energy–stress tensor for several matter fields has the same form as Eq. (1.94), for example, a null electromagnetic field, a massless scalar field, or a neutrino field (Kramer *et al.*, 1980; Stephani *et al.*, 2009). For the latter cases, however, the corresponding matter field equations must also be satisfied.

(γ) *An electromagnetic field*: For an electromagnetic field $F_{\mu\nu}$, the energy–stress tensor takes the form

$$T_{\mu\nu} = F_{\mu\lambda}F^{\lambda}{}_{\nu} - \frac{1}{4}g_{\mu\nu}F_{\rho\lambda}F^{\lambda\rho}, \tag{1.95}$$

where the antisymmetric tensor $F_{\mu\nu}$ satisfies the Maxwell equations

$$F_{[\mu\nu;\lambda]} = 0, \quad F_{\mu\nu;\lambda}g^{\nu\lambda} = 0. \tag{1.96}$$

Introducing the following notations (Newman and Penrose, 1962),

$$\Phi_0 \equiv F_{(0)(2)} = F_{\mu\nu}l^{\mu}m^{\nu}, \quad \Phi_2 \equiv -F_{(1)(3)} = -F_{\mu\nu}n^{\mu}\overline{m}^{\nu}, \tag{1.97a}$$

$$\Phi_1 \equiv \frac{1}{2}[F_{(0)(1)} - F_{(2)(3)}] = \frac{1}{2}(F_{\mu\nu}l^{\mu}n^{\nu} - F_{\mu\nu}m^{\mu}\overline{m}^{\nu}),$$

$$F_{\mu\nu} = 2\left\{-\Phi_0 n_{[\mu}\overline{m}_{\nu]} - \overline{\Phi}_0 n_{[\mu}m_{\nu]} + \Phi_2 l_{[\mu}m_{\nu]} + \overline{\Phi}_2 l_{[\mu}\overline{m}_{\nu]}\right\}$$
$$- 4\,\mathrm{Re}(\Phi_1)l_{[\mu}n_{\nu]} + 4i\,\mathrm{Im}(\Phi_1)m_{[\mu}\overline{m}_{\nu]}, \tag{1.97b}$$

we find that the Ricci scalars are given by

$$\Phi_{mn} = \Phi_m\overline{\Phi}_n, \quad \Lambda = 0, \quad (m, n = 0, 1, 2) \tag{1.98}$$

and that the Maxwell equations read

$$D\Phi_1 - \bar{\delta}\Phi_0 = (\pi - 2\alpha)\Phi_0 + 2\rho\Phi_1 - \kappa\Phi_2, \tag{1.99a}$$

$$D\Phi_2 - \bar{\delta}\Phi_1 = (\rho - 2\varepsilon)\Phi_2 + 2\pi\Phi_1 - \lambda\Phi_0, \tag{1.99b}$$

$$\delta\Phi_1 - \Delta\Phi_0 = (\mu - 2\gamma)\Phi_0 + 2\tau\Phi_1 - \sigma\Phi_2, \tag{1.99c}$$

$$\delta\Phi_2 - \Delta\Phi_1 = (\tau - 2\beta)\Phi_2 + 2\mu\Phi_1 - \nu\Phi_0. \tag{1.99d}$$

(δ) *A massless neutrino field*: In general, the case for a massless neutrino field is much more complicated than the previous ones. This is mainly due to the fact that a neutrino field is described by a two component spinor ϕ_A, which satisfies the neutrino Weyl equations

$$\sigma^\mu_{A\dot{B}}\phi^A{}_{;\mu} = 0, \tag{1.100}$$

where $\sigma^\mu_{A\dot{B}}$ are the complex Pauli spin matrices (Griffiths, 1980, 1991; Tsoubelis and Wang, 1991), and the spin indices A, B take the values $1, 2$. Then, the energy–stress tensor for a massless neutrino field takes the form

$$T_{\mu\nu} = i[\sigma_{\mu A\dot{B}}(\phi^A\phi^{\dot{B}}{}_{;\nu} - \phi^{\dot{B}}\phi^A{}_{;\nu}) + \sigma_{\nu A\dot{B}}(\phi^A\phi^{\dot{B}}{}_{;\mu} - \phi^{\dot{B}}\phi^A{}_{;\mu})]. \tag{1.101}$$

In a spinor basis (O_A, L_A), the neutrino spinor ϕ_A can be written as

$$\phi_A = \Phi O_A + \Psi L_A, \tag{1.102}$$

where O_A and L_A are normalized by the conditions

$$O_A L^A = -L_A O^A = 1. \tag{1.103}$$

In terms of Φ and Ψ and the spin coefficients, Eq. (1.98) takes the form (Griffiths, 1976a, 1991; Tsoubelis and Wang, 1991)

$$D\Phi + \bar{\delta}\Psi = (\rho - \varepsilon)\Phi + (\alpha - \pi)\Psi, \tag{1.104a}$$

$$\delta\Phi + \Delta\Psi = (\tau - \beta)\Phi + (\gamma - \mu)\Psi. \tag{1.104b}$$

The Ricci scalars are now given by

$$\Phi_{00} = i[\Psi D\overline{\Psi} - \overline{\Psi}D\Psi + \kappa\Phi\overline{\Psi} - \overline{\kappa}\Psi\overline{\Phi} + (\varepsilon - \bar{\varepsilon})\Psi\overline{\Psi}],$$

$$\Phi_{01} = \frac{i}{2}[\Psi\delta\overline{\Psi} - \overline{\Psi}\delta\Psi - \Psi D\overline{\Phi} + \overline{\Phi}D\Psi - (\bar{\rho} + \varepsilon + \bar{\varepsilon})\Psi\overline{\Phi} + (\beta - \bar{\alpha} - \pi)\Psi\overline{\Psi}$$
$$- \kappa\Phi\overline{\Phi} + \sigma\Phi\overline{\Psi}],$$

$$\Phi_{02} = -i[\Psi\delta\overline{\Phi} - \overline{\Phi}\delta\Psi + (\overline{\alpha} + \beta)\Psi\overline{\Phi} + \sigma\Phi\overline{\Phi} + \lambda\Psi\overline{\Psi}],$$

$$\Phi_{11} = \frac{i}{2}[\Phi D\overline{\Phi} - \overline{\Phi}D\Phi + \Psi\Delta\overline{\Psi} - \overline{\Psi}\Delta\Psi + (\overline{\varepsilon} - \varepsilon)\Phi\overline{\Phi} + (\tau + \overline{\pi})\overline{\Psi}\Phi$$
$$- (\overline{\tau} + \pi)\Psi\overline{\Phi} + (\gamma - \overline{\gamma})\Psi\overline{\Psi}],$$

$$\Phi_{12} = \frac{i}{2}[\Phi\delta\overline{\Phi} - \overline{\Phi}\delta\Phi - \Psi\Delta\overline{\Phi} + \overline{\Phi}\Delta\Psi + (\overline{\alpha} - \beta - \tau)\Phi\overline{\Phi} - (\mu + \gamma + \overline{\gamma})\Psi\overline{\Phi}$$
$$- \overline{\nu}\Psi\overline{\Psi} + \overline{\lambda}\Phi\overline{\Psi}],$$

$$\Phi_{22} = i[\Phi\Delta\overline{\Phi} - \overline{\Phi}\Delta\Phi + (\overline{\gamma} - \gamma)\Phi\overline{\Phi} + \overline{\nu}\Phi\overline{\Psi} - \nu\Psi\overline{\Phi}],$$

$$\Lambda = 0. \tag{1.105}$$

Equations (1.102) and (1.103) are the basic equations for a neutrino field.

(ε) *An isentropic perfect fluid*: The energy–stress tensor for a perfect fluid takes the form

$$T_{\mu\nu} = (\mu + p)u_\mu u_\nu - pg_{\mu\nu}, \quad u_\mu u_\nu g^{\mu\nu} = 1, \tag{1.106}$$

where u_μ is the four-velocity of the fluid, p the pressure, and μ the energy density (where μ must not be confused with the one used for spin coefficients). Substituting Eq. (1.106) into Eq. (1.21), we obtain

$$\mu_{;\nu}u^\nu + (\mu + p)u^\nu{}_{;\nu} = 0,$$
$$(\mu + p)u^\mu{}_{;\nu}u^\nu + (u^\mu u^\nu - g^{\mu\nu})p_{;\nu} = 0, \tag{1.107}$$

which are the conditions imposed on a perfect fluid. In order to completely describe a perfect fluid, however, Eq. (1.107) has to be supplemented by an equation of state (Taub, 1956, 1975, 1983). More frequently, the relation $p = p(\mu)$ is prescribed. We call a perfect fluid isentropic if the pressure p is a function of the energy density μ only. The simplest cases of the isentropic fluids are those with a "gamma equation of state"

$$p = (\gamma - 1)\mu, \tag{1.108}$$

where γ is a constant, and must not be confused with the one used for spin coefficients.

In general, matter fields need to satisfy some energy conditions. The energy conditions for a neutrino field are discussed by Griffiths (1980). These include the weak, dominant and strong energy conditions (Hawking and Eillis, 1973).

(a) *The weak energy condition*: This condition says that the energy density measured by any observer must be non-negative. Mathematically, it is equivalent to saying that for any time-like vector u_μ we must have

$$T^{\mu\nu} u_\mu u_\nu \geq 0. \tag{1.109}$$

Equation (1.109) is also true even for any null vector k_μ.

(b) *The dominant energy condition*: The dominant energy condition is stronger than the weak energy condition. Besides the requirement of Eq. (1.109), it also requires that for any observer the local energy flow vector $q^\mu \ (\equiv T^{\mu\nu} u_\mu)$ be non-space-like, i.e.

$$q^\mu q_\mu \geq 0. \tag{1.110}$$

(c) *The strong energy condition*: This is basically stated that the expansion of a time-like geodesic congruence with zero vorticity will monotonically decrease along a geodesic. Mathematically, this is equivalent to require $R_{\mu\nu} u^\mu u^\nu \geq 0$. Then, by the Einstein field equations, this implies that

$$T_{\mu\nu} u^\mu u^\nu \geq \left(\frac{1}{2} T - \frac{\Lambda}{\varkappa} \right) u^\nu u_\nu, \tag{1.111}$$

where u^μ denotes the tangential vector of the given time-like geodesic, and $T \equiv T^\nu{}_\nu$. It is normally said that the energy-momentum tensor $T_{\mu\nu}$ satisfies the strong energy condition, if it obeys Eq. (1.111) for $\Lambda = 0$ (Hawking and Eillis, 1973).

1.9. Spacetimes with Distribution-Valued Tensors

The study of singular surfaces in GR started with the pioneering work of Lanczos (1922, 1924). Since then, this problem has been widely investigated. In particular, in 1966 Israel first gave a complete analysis of the case in which there was a discontinuity of the first or second derivatives of the metric across a non-null (time-like or space-like) hypersurface (Israel, 1966, 1967). It was shown that the singular part of the energy–stress tensor is related to the jump of the second fundamental form across the singular hypersurface. If the jump is zero, we obtain the junction conditions first proposed by Darmois (1927). Later on, O'Brien and Synge (1952) and Lichnerowicz (1955; see also Papapetrou and Hamoufi, 1968) considered the same problem and proposed other junction conditions. However, Bonnor and Vichers (1981) showed that Darmois' junction conditions are equivalent to Lichnerowicz's, and implied by, but not equivalent to, O'Brien and Synge's.

The non-null surface case has attracted a lot of interest with various motivations since 1980s. In particular, the phase transitions that might have occurred in the early epoch of the universe resulted in the formation of topological defects, *domain walls, cosmic strings, monopoles and textures*, according to various field theories, including a wide variety of grand unified theories (GUT's) (Zel'dovish, Kobzarev and Okun, 1976; Kibble, 1976; Vilenkin, 1981). The time-varying gravitational fields associated with the formation of cosmic strings will create particles that have contributions to the average energy density of the Universe. These contributions could be especially important in the early Universe (Zel'dovish, 1980; Vilenkin, 1985; Xanthopoulos, 1986a, 1986b, 1987; Economou and Tsoubelis, 1988a, 1988b; Tsoubelis, 1989a; Letelier and Wang, 1995; Wang and Santos, 1996; Wang and Nogales, 1997; Nogales and Wang, 1998; Bronnikov, Santos and Wang, 2019). In particular, they were once believed to provide a mechanism to produce the cosmic microwave background (CMB) and large-scale structure observed in our Universe (Vilenkin and Shellard, 2000). However, later observations of CMB ruled out cosmic strings formed in the context of symmetry breaking in GUT's as the sources of the cosmological perturbations (Bennett *et al.*, 1996; Spergel *et al.*, 2007), and led to an upper bound (Ade *et al.*, 2016),

$$\frac{G\mu}{c^2} \lesssim 10^{-7}, \tag{1.112}$$

where μ denotes the string's tension.

Nevertheless, the subject has attracted recently lots of attention again in the framework of string/M-Theory, the so-called *cosmic superstrings* (Dvali and Vilenkin, 2004; Copeland, Myers and Polchinski, 2004), which were formed before inflation took place and stretched to macroscopic length scales in the inflationary phase. During the subsequent epochs, a complicated network of various string elements forms (Chernoff and Tye, 2018; and references therein). The main phenomenological consequence of a string network is the emission of GWs (Ringeval and Suyama, 2017; and references therein), generating bursts at cusps, kinks and junctions, as well as a stochastic gravitational wave background. Low tension strings are natural in string/M-theory, and can easily satisfy the observational bounds given above (Chernoff, Flanagan and Wardell, 2018).

On the other hand, domain walls played an important role in the theory of inflation (Linde, 1984; Ipser and Sikivie, 1984; Brandenberger, 1985;

Laguna-Castillo and Matzner, 1986; Berezin, Kuzmin and Tkachev, 1987; Hill, Schramm and Fry, 1989; Wang, 1991c, 1991d, 1991e, 1992a, 1992b, 1992c, 1992d, 1992e, 1992f, 1993, 1994; Schmidt and Wang, 1993; Letelier and Wang, 1993a, 1993b, 1993c; Khorrami and Mansouri, 1994; Letelier and Wang, 1995; Wang and Letelier, 1995a, 1995b; Paiva and Wang, 1995). Later, to solve the long-standing hierarchy problem, brane world scenarios were proposed in the late 1990s (Arkani-Hamed, Dimopoulos and Dvali, 1998, 1999; Randall and Sundrum, 1999a, 1999b), and have been intensively studied since then. As a matter of fact, the field has been so extensively investigated that it is very difficult to provide a list of references, so here we simply refer readers to the review articles of the field (Maartens, 2004; Wands, 2006; Maartens and Koyama, 2010). However, most of these works have been phenomenological in nature, although there have also been studies in which the models are built in the framework of string/M-theory (Horava and Witten, 1995, 1996; Lukas *et al.*, 1999; Goldberger and Wise, 1999; Kachru, Schulz and Trivedi, 2003; Braun and Ovrut, 2006; Gray, Lukas and Ovrut, 2007; Devin *et al.*, 2009; Wu, Gong and Wang, 2009; Wang, 2010; Wang and Santos, 2010). Another important application of these string/M-theory inspired models is to provide a mechanism to produce the late cosmic acceleration of the Universe without provoking the presence of dark energy (Roy, 2003; Townsend and Wohlfarth, 2003; Wohlfarth, 2003; Gong, Wang and Wu, 2008; Wang and Santos, 2008; Wu *et al.*, 2008; Wu, Gong and Wang, 2009; Devin *et al.*, 2009).

Gravitational collapse and formation of spacetime singularities is another area in which time-like thin shells have been heavily used to model collapsing matter sources (Barrabés, Israel and Letelier, 1991; Apostolatos and Thorne, 1992; Barrabés *et al.*, 1992; Echeverria 1993; Letelier and Wang, 1994; Holvorcem, Letelier and Wang, 1995; Wang and Letelier, 1995c; Wang and de Oliveira, 1997; Pereira and Wang, 2000a, 2000b; Wu *et al.*, 2003; Herrera and Santos, 2005; Nakao *et al.*, 2007; Rocha *et al.*, 2008a, 2008b; Tziolas and Wang, 2008; Tziolas, Wang and Wu, 2009; Nakao *et al.*, 2009; Chan *et al.*, 2011; Sharma *et al.*, 2011). As a matter of fact, this used to be one of the main motivations to study singular surfaces in the early time of GR.

In the case where the surface of discontinuity is null, the concept of the second fundamental form becomes invalid, and the methods used for the non-null surface case cannot be used. Dautcourt (1964) and Penrose (1972) first studied the null case, and the latter used spinor techniques. The null case is interesting, because it relates to gravitational shock and impulsive waves, and to null dust shells consisting of massless particles (Wang and de Oliveira, 1997; Pereira and Wang, 2002; Barrabés and Hogan, 2003). Specially, it relates to the collision and interaction of plane gravitational waves and matter (Griffiths, 1991), which is one of the main subjects of this book.

In order to give a unified description for the above two cases, Taub (1980) introduced the formalism of distribution theory (see also Pantaleo, 1979; Greenwald *et al.*, 2013). Later on, Clarke and Dray (1987) and Barrabés (1989) generalized the notions of the first and second fundamental forms to arbitrary surfaces.

In review of the wide range of applications of singular surfaces mentioned above, in this section we shall provide a general and systematical description, which is applicable for both non-null and null surfaces. Although such a presentation is restricted only to GR, its generalizations to other theories of gravity are straightforward.

In the rest of the section, we are mainly concerned with spacetimes whose curvature tensors contain Dirac delta functions with supports on submanifolds or even at isolated events (for more general cases, see, for example, Greenwald *et al.*, 2013). Since the Riemann curvature tensor includes the second derivatives of the metric and is quadratic in the first derivatives, if we require the metric to be continuous and the first and second derivatives of the metric to have finite jumps across the singular surface, then the Riemann curvature tensor will contain Dirac delta functions.

To begin with, we assume that we are given a spacetime manifold $(\Omega, g_{\mu\nu})$ and a surface Σ, described by a function $\varphi(x^\mu)$ as

$$\Sigma = \{x^\mu : \varphi(x^\mu) = 0\}. \tag{1.113}$$

The surface Σ divides Ω into two open regions, $\Omega^+ = \{x^\mu : \varphi(x^\mu) > 0\}$ and $\Omega^- = \{x^\mu : \varphi(x^\mu) < 0\}$. We assume that: (i) the restrictions $g_{\mu\nu}^\pm \equiv g_{\mu\nu}|_{\Omega^\pm}$ are at least C^3 and (ii) $g_{\mu\nu}$ is C^0 across Σ.

Note that for the sake of convenience in the following we consider only the cases in which the spacetime has only one singular surface, but the extension to the cases with multiple singular surfaces is straightforward.

The normal vector to Σ is defined as

$$k_\mu \equiv \frac{\partial \varphi}{\partial x^\mu} = \varphi_{;\mu}. \tag{1.114}$$

We define a step function $\theta(\varphi)$ on Ω by[4]

$$\theta(\varphi) = \begin{cases} 1 & \varphi > 0, \\ \dfrac{1}{2} & \varphi = 0, \\ 0 & \varphi < 0. \end{cases} \qquad (1.115)$$

Then, we have

$$\frac{\partial \theta(\varphi)}{\partial x^\mu} = \delta(\varphi) k_\mu, \qquad (1.116)$$

where $\delta(\varphi)$ denotes the Dirac delta function with support on Σ. Thus, for a test function f of compact support, we have

$$\langle \delta(\varphi), f \rangle \equiv \int_\Omega \sqrt{-g}\, \delta(\varphi) f dV = \int_{\partial \Omega^-} f dS = - \int_{\partial \Omega^+} f dS, \qquad (1.117)$$

where dS is the invariant volume element induced on the hypersurface Σ.

The nth derivative $\delta^{(n)}(\varphi)$ $(n \geq 1)$ of $\delta(\varphi)$ is defined in a standard way,

$$\langle \delta^{(n)}(\varphi), f \rangle = - \left\langle \delta^{(n-1)}(\varphi), \frac{\partial f}{\partial \varphi} \right\rangle, \qquad (1.118)$$

and the following relations are valid (Gelfand and Shilov, 1964):

$$\frac{\partial H(\varphi)}{\partial x^\lambda} = \frac{\partial \varphi}{\partial x^\lambda} \delta(\varphi), \quad \frac{\partial \delta^{(n)}(\varphi)}{\partial x^\lambda} = \frac{\partial \varphi}{\partial x^\lambda} \delta^{(n+1)}(\varphi), \quad (n = 0, 1, 2, \ldots),$$

$$\varphi \delta^{(n)}(\varphi) = -n\delta^{(n-1)}(\varphi), \quad \varphi^n \delta^{(n)}(\varphi) = (-1)^n n! \delta(\varphi), \quad (n = 1, 2, \ldots).$$

$$(1.119)$$

In addition, if $\mathcal{F}$ is a function defined in a neighborhood of Σ, we define the distribution $\mathcal{F}\delta^{(n)}(\varphi)$ by letting it act on a test function f (Greenwald *et al.*, 2013) as follows:

$$\langle \mathcal{F}\delta^{(n)}(\varphi), f \rangle = \langle \delta^{(n)}(\varphi), f\mathcal{F} \rangle. \qquad (1.120)$$

---

[4]Note that instead of defining a step function as Eq. (1.115), we can use the Heaviside step function, $H(\varphi)$, which is unity for the non-negative arguments and otherwise zero. But all the following results are valid for both of them, if we just simply replace one by another. In fact, the specific value of $\theta(\varphi)$ at $\varphi = 0$ is irrelevant in the sense of distributions, and does not affect the following results at all.

Then, we can see that the product $\mathcal{F}\delta(\varphi)$ is well defined whenever $\mathcal{F}$ is C^0, and it depends only on the restriction $\mathcal{F}|_\Sigma$ of $\mathcal{F}$ to Σ. More generally, the product $\mathcal{F}\delta^{(n)}(\varphi)$ is well defined, provided that $\mathcal{F}$ is C^n. Clearly, such defined $\mathcal{F}\delta^{(n)}(\varphi)$ depends only on the values of $\mathcal{F}$ and its partial derivatives of the mth order evaluated on Σ, where $m \leq n$.

By using the continuity of $g_{\mu\nu}$, we can write $g_{\mu\nu}$ as

$$g_{\mu\nu} = \theta(\varphi)g_{\mu\nu}^+ + (1 - \theta(\varphi))g_{\mu\nu}^-. \tag{1.121}$$

Then, for a test function f we find that the distribution derivative of $g_{\mu\nu}$ is

$$\begin{aligned}
\langle \partial_\lambda g_{\mu\nu}, f \rangle &= \langle \partial_\lambda \left\{ \theta(\varphi)g_{\mu\nu}^+ + (1 - \theta(\varphi))g_{\mu\nu}^- \right\}, f \rangle \\
&= \langle \{ \theta(\varphi)g_{\mu\nu,\lambda}^+ + (1 - \theta(\varphi))g_{\mu\nu,\lambda}^- \}, f \rangle \\
&\quad + \langle \{ g_{\mu\nu}^+ \partial_\lambda \theta(\varphi) + g_{\mu\nu}^- \partial_\lambda (1 - \theta(\varphi)) \}, f \rangle. \tag{1.122}
\end{aligned}$$

Since

$$\begin{aligned}
&\langle \{ g_{\mu\nu}^+ \partial_\lambda \theta(\varphi) + g_{\mu\nu}^- \partial_\lambda (1 - \theta(\varphi)) \}, f \rangle \\
&= \langle \nabla_\lambda \theta(\varphi), g_{\mu\nu}^+ f \rangle + \langle \nabla_\lambda (1 - \theta(\varphi)), g_{\mu\nu}^- f \rangle \\
&= - \int_{\Omega^+} (\nabla_\lambda (g_{\mu\nu}^+ f)) \, dV - \int_{\Omega^-} (\nabla_\lambda (g_{\mu\nu}^- f)) \, dV \\
&= - \int_{\partial\Omega^+} (g_{\mu\nu}^+ f) \, dS_\lambda - \int_{\partial\Omega^-} (g_{\mu\nu}^- f) \, dS_\lambda \\
&= - \int_{\partial\Omega^+} (g_{\mu\nu}^+ - g_{\mu\nu}^-) \, f \, dS_\lambda = 0, \tag{1.123}
\end{aligned}$$

we have

$$\langle \partial_\lambda g_{\mu\nu}, f \rangle = \langle \{ \theta(\varphi)g_{\mu\nu,\lambda}^+ + (1 - \theta(\varphi))g_{\mu\nu,\lambda}^- \}, f \rangle. \tag{1.124}$$

Equation (1.123) can be simply written as

$$g_{\mu\nu,\lambda} = \theta(\varphi)g_{\mu\nu,\lambda}^+ + (1 - \theta(\varphi))g_{\mu\nu,\lambda}^-. \tag{1.125}$$

On the other hand, from the definition of the connection coefficients $\Gamma_{\nu\lambda}^\mu$ we find that

$$\Gamma_{\nu\lambda}^\mu = \theta(\varphi)\Gamma_{\nu\lambda}^{+\mu} + (1 - \theta(\varphi))\Gamma_{\nu\lambda}^{-\mu}, \tag{1.126}$$

where the superscripts "$\pm$" always refer to the quantities calculated in $\Omega^\pm$. If T is a vector field in Ω, and in addition if T and its derivatives have finite discontinuities across Σ, we define distributions as follows (Taub, 1980):

$$(T^\mu)^D \equiv \theta(\varphi)T^{+\mu} + (1 - \theta(\varphi))T^{-\mu}, \tag{1.127a}$$

$$\left(T^\mu{}_{;\nu}\right)^D \equiv \theta(\varphi)T^{+\mu}{}_{;\nu} + (1 - \theta(\varphi))T^{-\mu}{}_{;\nu}. \tag{1.127b}$$

For distribution-valued vector and tensor fields, we define the covariant differentiation as

$$[(T^\mu)^D]_{;\nu} \equiv [(T^\mu)^D]_{,\nu} + (T^\lambda)^D \Gamma^\mu_{\nu\lambda}, \tag{1.128}$$

where $\Gamma^\mu_{\nu\lambda}$ is given by Eq. (1.126).

If we define the symbol $[F]^-$ to be the discontinuity in the function F at Σ, i.e.

$$[F]^- = \lim_{x \to y^+} F - \lim_{x \to y^-} F, \quad y \in \Sigma, \tag{1.129}$$

where $y^\pm$ indicate the limits to be taken in $\Omega^\pm$, respectively, we find

$$[(T^\mu)^D]_{;\nu} = \left(T^\mu{}_{;\nu}\right)^D + [T^\mu]^- k_\nu \delta(\varphi) - [T^\lambda]^- [\Gamma^\mu_{\nu\lambda}]^- \theta(\varphi)(1 - \theta(\varphi)). \tag{1.130}$$

It must be noted that, since for any given test function f, we have

$$\langle \theta(\varphi)(1 - \theta(\varphi)), f \rangle = \int_\Omega [\theta(\varphi)(1 - \theta(\varphi))f] dV$$

$$= \int_{\Omega^+} (1 - \theta(\varphi)) f dV = 0, \tag{1.131}$$

so we can set

$$\theta(\varphi)(1 - \theta(\varphi)) = 0, \tag{1.132}$$

in the sense of distributions. However, in what follows we prefer to keep such terms (Taub, 1980).

On the other hand, in the neighborhood of Σ, the metric tensor $g^\pm{}_{\mu\nu}$ can be written as

$$g^\pm{}_{\mu\nu} = g^\pm{}_{\mu\nu} \left(\varphi\left(x^\mu\right), x^\mu\right). \tag{1.133}$$

Thus, we have

$$g^\pm_{\mu\nu} = g^0_{\mu\nu} + \left.\frac{\partial g^\pm_{\mu\nu}}{\partial \varphi}\right|_{\varphi=0} \varphi + \frac{1}{2} \left.\frac{\partial^2 g^\pm_{\mu\nu}}{\partial \varphi^2}\right|_{\varphi=0} \varphi^2 + \mathcal{O}\left(\varphi^3\right). \tag{1.134}$$

Then, it is easy to show that

$$
\begin{aligned}
[g_{\mu\nu,\lambda}]^- &= k_\lambda \gamma_{\mu\nu}, \\
[g_{\mu\nu,\lambda\delta}]^- &= k_{\lambda,\delta}\gamma_{\mu\nu} + k_\lambda \gamma_{\mu\nu,\delta} + k_\delta \gamma_{\mu\nu,\lambda} + k_\lambda k_\delta \hat{\gamma}_{\mu\nu},
\end{aligned}
\tag{1.135}
$$

where $\gamma_{\mu\nu} \equiv [g_{\mu\nu,\varphi}]^-$ and $\hat{\gamma}_{\mu\nu} \equiv [g_{\mu\nu,\varphi\varphi}]^-$. Hence, we have

$$
[\Gamma^\mu_{\nu\lambda}]^- = \frac{1}{2}[k_\nu \gamma^\mu{}_\lambda + k_\lambda \gamma^\mu{}_\nu - k^\mu \gamma_{\nu\lambda}].
\tag{1.136}
$$

From the definition of the Riemann tensor given by Eq. (1.13), on the other hand, we find

$$
\begin{aligned}
R^\sigma{}_{\mu\nu\lambda} &= \theta(\varphi)R^{+\sigma}{}_{\mu\nu\lambda} + (1 - \theta(\varphi))R^{-\sigma}{}_{\mu\nu\lambda} + \delta(\varphi)H^\sigma{}_{\mu\nu\lambda} \\
&\quad + \theta(\varphi)(1 - \theta(\varphi))I^\sigma{}_{\mu\nu\lambda},
\end{aligned}
\tag{1.137}
$$

where

$$
\begin{aligned}
H^\sigma{}_{\mu\nu\lambda} &\equiv k_\nu [\Gamma^\sigma_{\mu\lambda}]^- - k_\lambda [\Gamma^\sigma_{\mu\nu}]^-, \\
I^\sigma{}_{\mu\nu\lambda} &\equiv [\Gamma^\sigma_{\lambda\delta}]^- [\Gamma^\delta_{\mu\nu}]^- - [\Gamma^\sigma_{\nu\delta}]^- [\Gamma^\delta_{\mu\lambda}]^-,
\end{aligned}
\tag{1.138}
$$

and $R^{\pm\sigma}{}_{\mu\nu\lambda}$ are the Riemann tensor calculated, respectively, in $\Omega^\pm$. Then, the Ricci tensor is given by

$$
R_{\mu\nu} = \theta(\varphi)R^+{}_{\mu\nu} + (1 - \theta(\varphi))R^-{}_{\mu\nu} + \delta(\varphi)H_{\mu\nu} + \theta(\varphi)(1 - \theta(\varphi))I_{\mu\nu},
\tag{1.139}
$$

where

$$
R^\pm{}_{\mu\nu} \equiv R^{\pm\sigma}{}_{\mu\sigma\nu},
$$

$$
H_{\mu\nu} \equiv H^\sigma{}_{\mu\sigma\nu} = \frac{1}{2}[k_\mu k_\lambda \gamma^\lambda{}_\nu + k_\nu k_\lambda \gamma^\lambda{}_\mu - k^\lambda k_\lambda \gamma_{\mu\nu} - k_\mu k_\nu \gamma^\lambda{}_\lambda],
$$

$$
\begin{aligned}
I_{\mu\nu} \equiv I^\sigma{}_{\mu\sigma\nu} = \frac{1}{4}\Big[&k_\mu k_\nu \gamma^{\lambda\delta}\gamma_{\lambda\delta} + 2k_\lambda k_\delta \gamma^\lambda{}_\mu \gamma^\delta{}_\nu - (k_\mu \gamma^\lambda{}_\nu + k_\nu \gamma^\lambda{}_\mu)k_\lambda \gamma^\delta{}_\delta \\
&- (2\gamma_{\mu\delta}\gamma^\delta{}_\nu - \gamma^\delta{}_\delta \gamma_{\mu\nu})k_\lambda k^\lambda \Big].
\end{aligned}
\tag{1.140}
$$

The Ricci scalar is given by

$$
R \equiv R^\nu{}_\nu = \theta(\varphi)R^+ + (1 - \theta(\varphi))R^- + \delta(\varphi)H + \theta(\varphi)(1 - \theta(\varphi))I,
\tag{1.141}
$$

where

$$R^{\pm} \equiv R^{\pm\nu}{}_{\nu},$$

$$H \equiv H^{\nu}{}_{\nu} = k_{\mu}k_{\nu}\gamma^{\mu\nu} - k^{\nu}k_{\nu}\gamma^{\mu}{}_{\mu},$$

$$I \equiv I^{\nu}{}_{\nu} = \frac{1}{2}\left(\gamma_{\mu\delta}\gamma^{\delta}{}_{\nu} - \gamma^{\delta}{}_{\delta}\gamma_{\mu\nu}\right)k^{\mu}k^{\nu} + \frac{1}{4}\left(\gamma_{\mu}{}^{\mu}\gamma^{\delta}{}_{\delta} - \gamma^{\delta\mu}\gamma_{\mu\delta}\right)k_{\nu}k^{\nu}. \tag{1.142}$$

Now, we are ready to write down the generalized Einstein field equations, which take the form,

$$R_{\mu\nu} - \frac{1}{2}g_{\mu\nu}R = G^{D}{}_{\mu\nu} + \delta(\phi)\left(H_{\mu\nu} - \frac{1}{2}g_{\mu\nu}H\right)$$

$$+ \theta(\varphi)(1 - \theta(\varphi))\left(I_{\mu\nu} - \frac{1}{2}g_{\mu\nu}I\right)$$

$$= T^{D}_{\mu\nu} + \delta(\varphi)\tau_{\mu\nu} + \theta(\varphi)(1 - \theta(\varphi))J_{\mu\nu}, \tag{1.143}$$

where

$$G^{D}{}_{\mu\nu} \equiv \theta(\varphi)G^{+}{}_{\mu\nu} + (1 - \theta(\varphi))G^{-}{}_{\mu\nu},$$

$$T^{D}{}_{\mu\nu} \equiv \theta(\varphi)T^{+}{}_{\mu\nu} + (1 - \theta(\varphi))T^{-}{}_{\mu\nu}, \tag{1.144}$$

and $\tau_{\mu\nu}$ and $J_{\mu\nu}$ denote the energy–stress tensors with supports only on the surface Σ. Equation (1.143) can be written as

$$G^{\pm}{}_{\mu\nu} = T^{\pm}{}_{\mu\nu}, \tag{1.145a}$$

$$H_{\mu\nu} - \frac{1}{2}g_{\mu\nu}H = \tau_{\mu\nu}, \tag{1.145b}$$

$$I_{\mu\nu} - \frac{1}{2}g_{\mu\nu}I = J_{\mu\nu}. \tag{1.145c}$$

From Eq. (1.145b), we find

$$\tau_{\mu\nu} = \frac{1}{2}[\gamma^{\delta}{}_{\delta}\left(k^{\lambda}k_{\lambda}g_{\mu\nu} - k_{\mu}k_{\nu}\right) + \left(k_{\mu}\gamma^{\lambda}{}_{\nu} + k_{\nu}\gamma^{\lambda}{}_{\mu}\right)k_{\lambda}$$

$$- k^{\lambda}k_{\lambda}\gamma_{\mu\nu} - g_{\mu\nu}k_{\delta}k_{\lambda}\gamma^{\delta\lambda}], \tag{1.146}$$

which is applicable for any kind of thin shells. Contracting it with k^{ν} we obtain

$$\tau_{\mu\nu}k^{\nu} = 0. \tag{1.147}$$

In the case of a null shell, Eq. (1.146) reduces to

$$\tau_{\mu\nu} = \frac{1}{2}[(k_{\mu}\gamma^{\lambda}{}_{\nu} + k_{\nu}\gamma^{\lambda}{}_{\mu})k_{\lambda} - \gamma^{\delta}{}_{\delta}k_{\mu}k_{\nu} - g_{\mu\nu}k_{\delta}k_{\lambda}\gamma^{\delta\lambda}]. \tag{1.148}$$

And it is easy to show that in this case $\tau_{\mu\nu}$ is traceless, $\tau_\nu{}^\nu = 0$.

On the other hand, from Eq. (1.128) it can be shown that the Bianchi identities take the form

$$R^\sigma{}_{\mu\nu\lambda;\rho} + R^\sigma{}_{\mu\rho\nu;\lambda} + R^\sigma{}_{\mu\lambda\rho;\nu} = \theta(\varphi)(1 - \theta(\varphi))A^\sigma{}_{\mu\nu\lambda\rho}, \qquad (1.149)$$

where $A^\sigma{}_{\mu\nu\lambda\rho}$ is a tensor defined on Σ. Thus, the right-hand side of Eq. (1.149) vanishes everywhere except on Σ, and

$$\langle T_\sigma{}^{\mu\nu\lambda\rho}, \theta(\varphi)(1 - \theta(\varphi))A^\sigma{}_{\mu\nu\lambda\rho}\rangle = 0, \qquad (1.150)$$

for any C^1 tensor $T_\sigma{}^{\mu\nu\lambda\rho}$ of compact support in Ω.

Combining Eqs. (1.137) and (1.149) we find

$$\left(R^{\mu\nu} - \frac{1}{2}g^{\mu\nu}R\right)_{;\nu} = \theta(\varphi)(1 - \theta(\varphi))d^\mu, \qquad (1.151)$$

where d^μ is a vector defined on Σ. Thus, Eq. (1.149) implies that

$$\left\langle T_\mu, \left(R^{\mu\nu} - \frac{1}{2}g^{\mu\nu}R\right)_{;\nu}\right\rangle = 0, \qquad (1.152)$$

for any C^1 vector T_μ of compact support. As a consequence of Eqs. (1.143), (1.147) and (1.151), we obtain

$$\left(R^{\mu\nu} - \frac{1}{2}g^{\mu\nu R}\right)_{;\nu} = \theta(\varphi)T^{+\mu\nu}{}_{;\nu} + (1 - \theta(\varphi))T^{-\mu\nu}{}_{;\nu}$$

$$+ \delta(\varphi)(\tau^{\mu\nu}{}_{;\nu} + [T^{\mu\nu}]^- k_\nu) + \theta(\varphi)(1 - \theta(\varphi))J^{\mu\nu}{}_{;\nu}$$

$$= \theta(\varphi)(1 - \theta(\varphi))d^\mu, \qquad (1.153)$$

or equivalently

$$T^{\pm\mu\nu}{}_{;\nu} = 0, \qquad (1.154a)$$

$$\tau^{\mu\nu}{}_{;\nu} = -[T^{\mu\nu}]^- k_\nu, \qquad (1.154b)$$

$$J^{\mu\nu}{}_{;\nu} = d^\mu, \qquad (1.154c)$$

which are the generalized equations for the conservation of energy and stress of the sources.

Chapter 2

Plane Gravitational Waves

In this chapter, we investigate the plane gravitational wave, which is distinguished by both its intuitive physical significance and mathematical simplicity. We shall show how this class of null fields fits into the more general frame of pure radiation fields. In particular, in Section 2.1, we give a general description of null fields, and then in Section 2.2 we restrict ourselves to plane gravitational waves. In Section 2.3, as an example, we consider the Aichelburg–Sexl solution (Aichelburg and Sexl, 1971), which represents the gravitational wave produced by a massless particle. The solution was obtained by first making a Lorentz boost to the spherically symmetric Schwarzschild vacuum solution and then taking the massless limit. In Section 2.4, we study the polarization of a plane gravitational wave and express the polarization angle explicitly in terms of the Weyl scalars. Finally, in Section 2.5, we study the singularities of the gravitational plane wave spacetimes, and show explicitly that all such spacetimes are physically singular at the focusing surface, except for only two particular cases, in which the distortions of a freely falling observer remains finite when approaching this focusing point (Wang *et al.*, 2018). Its relevance to the gravitational memory effects (Favata, 2010; Bieri, Garfinkle and Yunes, 2017) and "soft-graviton" theorems (Hawking, Perry and Strominger, 2016, 2017; Strominger, 2017) is discussed.

2.1. Null Fields

A plane gravitational wave is a *null field*. To study null fields, we first review the general properties of a Petrov-type N field (Petrov, 1955), since the null fields belong to the latter.

Theorem 2.1 (Ehlers and Kundt, 1962). *In a Petrov-type N spacetime, there exists an orthogonal tetrad $\lambda^\mu_{(\alpha)}$ so that the Weyl tensor takes the form*

$$C_{\mu\nu\lambda\sigma} = m^{(1)}_{\mu\nu}m^{(1)}_{\lambda\sigma} - m^{(2)}_{\mu\nu}m^{(2)}_{\lambda\sigma}, \tag{2.1}$$

where

$$
\begin{aligned}
m^{(1)}_{\mu\nu} &\equiv 2k_{[\mu}\lambda_{(2)\nu]}, \quad m^{(2)}_{\mu\nu} \equiv 2k_{[\mu}\lambda_{(3)\nu]}, \\
k_\mu &\equiv \lambda_{(0)\nu} + \lambda_{(1)\nu}, \quad k_\mu k^\mu = k_\mu\lambda^\mu_{(2)} = k_\mu\lambda^\mu_{(3)} = 0.
\end{aligned} \tag{2.2}
$$

A null field is a Petrov-type N vacuum field. Thus, according to Theorem 2.1, we have

$$R_{\mu\nu\lambda\sigma} = C_{\mu\nu\lambda\sigma} = m^{(1)}_{\mu\nu}m^{(1)}_{\lambda\sigma} - m^{(2)}_{\mu\nu}m^{(2)}_{\lambda\sigma}, \tag{2.3}$$

for a null field.

Further properties of a null field can be summarized in the following theorems.

Theorem 2.2 (Ehlers and Kundt, 1962). *The null vector k_μ, uniquely determined (up to a sign) by Eq. (2.3), forms a congruence of shear-free geodesics,*

$$\sigma = 0, \tag{2.4}$$

and satisfies the relations

$$R^\mu_{\ \nu\lambda\sigma}k_\mu = 0. \tag{2.5}$$

A null field is physically interpreted as a pure radiation field in analogy to the electromagnetic pure radiation fields. This interpretation has been further justified by the study of asymptotic behaviors of the fields at large distance from sources (Sachs, 1960–1962).

To see the physical meaning of such a null field, let us consider the geodesic deviations. Suppose that an observer is moving in such a null field by following a time-like geodesic with his four-velocity u_μ. Without loss of generality, we assume that the observer moves perpendicular to the two-dimensional surface spanned by $\lambda^\mu_{(2)}$ and $\lambda^\mu_{(3)}$, i.e. $\lambda^\mu_{(2)}u_\mu = \lambda^\mu_{(3)}u_\mu = 0$. Substituting Eq. (2.3) into Eq. (1.29) and replacing t^μ by u^μ, we obtain

$$\frac{D^2\eta^\mu}{D\tau^2} = -(k_\lambda u^\lambda)^2[\lambda^\mu_{(2)}\lambda^\nu_{(2)} - \lambda^\mu_{(3)}\lambda^\nu_{(3)}]\eta_\nu, \tag{2.6}$$

where τ is the proper time used by the observer. Recalling the discussions presented in Sections 1.4 and 1.7, we can see that a circle of relative accelerations (with respect to the observer) goes over into an ellipse, with the minor axis along $\lambda_{(2)}^{\mu}$. Taking into account the direction of relative acceleration, we find that $\lambda_{(a)}^{\mu}$ are characterized as eigendirections (Ehlers and Kundt, 1962). We call $\lambda_{(2)}^{\mu}$ the direction of polarization with respect to u^{μ} (Ehlers and Kundt, 1962). The spatial projection of k^{μ} is orthogonal to the plane of relative accelerations, and $(u^{\mu}k_{\mu})^2$ is the magnitude of relative accelerations for neighboring test particles. That is, k^{μ} determined by Eq. (2.3) describes the direction of the propagation of the null field, and $(u^{\mu}k_{\mu})^2$ represents the strength of the null field (measured by the observer u^{μ}). Thus, considering Eq. (1.88) we have the following theorem.

Theorem 2.3 (Ehlers and Kundt, 1962). *Null fields are characterized as purely transverse vacuum fields. There exists a null vector k^{μ} such that relative accelerations and relative rotations of inertial directions are orthogonal to the spatial projection of k^{μ}.*

2.2. Plane Gravitational Waves

A null field with a non-expanding and non-rotating null geodesic congruence (ray) is called a *plane-fronted gravitational wave*, or in short, a pp-wave. The term "plane-fronted" is characterized by the vanishing of expansion and rotation of the null geodesics.

According to the definition of a pp-wave and Theorem 2.2, we find that the expansion, rotation and shear of a pp-wave are zero

$$\theta_{pp} = \omega_{pp} = \sigma_{pp} = 0. \tag{2.7}$$

Theorem 2.4 (Ehlers and Kundt, 1962). *In a pp-wave spacetime, there exists a covariant constant vector $\hat{k}^{\mu}$, which is collinear with the ray vector k^{μ} defined by, Eq. (2.3),*

$$k^{\mu} = k^{1/2}\hat{k}^{\mu}. \tag{2.8}$$

The above theorem implies that in a pp-wave spacetime the null geodesics formed by $\hat{k}^{\mu}$ (or k^{μ}) are parallel to each other. Introducing a function $\hat{u}$ such that

$$\hat{k}_{\mu} = \hat{u}_{,\mu}, \tag{2.9}$$

then the metric for a pp-wave can be written in the following form (Ehlers and Kundt, 1962)

$$ds^2 = 2d\hat{u}dZ + 2\,\mathrm{Re}(f)(d\hat{u})^2 - (dX)^2 - (dY)^2, \tag{2.10}$$

where $\{x^\mu\} = \{\hat{u}, Z, X, Y\}$ and $f = f(X, Y, \hat{u})$ is a complex function analytic with respect to X and Y and satisfies the following conditions:

$$f_{,\zeta\zeta} = ke^{i\theta}, \quad \zeta \equiv X + iY, \tag{2.11}$$

where $f_{,\zeta} \equiv \partial f/\partial\zeta$, and so on. The functions k and θ are real. Following the discussions carried out in Section 2.1, we can show that for the metric (2.10) the function k is the amplitude of the pp-wave, and θ the polarization angle, relative to the dX-axis. When θ is constant, we say the pp-wave is linearly polarized.

A pp-wave is called a *plane gravitational wave*, when k and θ are the functions of $\hat{u}$ only. Thus, a plane gravitational wave is a pp-wave with constant amplitude and polarization in every wavefront. From Eq. (2.11) we find that for a plane gravitational wave the function f satisfies

$$f_{,\zeta\zeta\zeta} = 0. \tag{2.12}$$

Then, it can be shown that the function f now takes the form,

$$\text{Re}(f) = (X^2 - Y^2)H_+(\hat{u}) + 2XYH_\times(\hat{u}), \tag{2.13}$$

where the functions $H_+(\hat{u})$ and $H_\times(\hat{u})$ determine the type of a plane gravitational wave, for example, when $H_\times(\hat{u}) = 0$, the plane gravitational wave is linearly (or collinearly) polarized. In addition, if $H_+(\hat{u}) = H(\hat{u})$, where $H(\hat{u})$ is the Heaviside step function, the wave is a linearly polarized shock wave, and if $H_+(\hat{u}) = \delta(\hat{u})$, it is an impulsive wave (Penrose, 1968). The linearly polarized plane gravitational waves were first studied by Brinkman (1923) and their physical interpretations were presented by Robinson (Ehlers and Kundt, 1962). Later on, Rosen (1937) studied the same problem but by using a different form of the metric. The general case was studied by Bondi (1957), Bondi, Pirani and Robinson (1959), and Jordan, Ehlers and Kundt (1960). The global properties for a plane gravitational wave spacetime was not studied until Penrose (1965) who first discussed the focusing effects. Later, Belinsky (1980), Carr and Verdaguer (1984), Ibañez and Verdaguer (1983, 1986), and Verdaguer (1987) studied the plane gravitational waves by using soliton technique (Belinsky and Verdaguer, 2001), developed by Belinsky and Zakharov (1978, 1979). See also Bondi and Pirani (1989).

2.3. Aichelburg–Sexl Plane-Fronted Gravitational Wave

It is well known that the spherically symmetric Schwarzschild vacuum solution represents a gravitational field produced by a point-like particle with mass m. In the isotropic coordinates, it is given by (D'Inverno, 2003),

$$ds^2 = \frac{(1-A)^2}{(1+A)^2}dt^2 - (1+A)^4(dx^2 + dy^2 + dz^2), \qquad (2.14)$$

with $A \equiv m/(2r)$, where $r \equiv \sqrt{x^2 + y^2 + z^2}$. Starting with this form of metric, Aichelburg and Sexl (1971) were able to obtain the gravitational field produced by a massless particle, after first making a Lorentz boost and then taking the massless limit. In doing so, they were able to show that such an obtained solution takes exactly the same form as that given by Eq. (2.10) for a plane-fronted gravitational wave.

To show the above claim, let us first consider the Lorentz boost along the x-direction,

$$\bar{t} = \gamma(t + vx), \quad \bar{x} = \gamma(x + vt), \quad \bar{y} = y, \quad \bar{z} = z, \qquad (2.15)$$

where $\gamma[\equiv 1/(1-v^2)^{1/2}]$ is the Lorentz factor. Then, the above metric takes the form,

$$ds^2 = (1+A)^2(d\bar{t}^2 - d\bar{x}^2 - d\bar{y}^2 - d\bar{z}^2)$$
$$- \gamma^2 \left[(1+A)^4 - \left(\frac{1-A}{1+A}\right)^2\right](d\bar{t} - vd\bar{x})^2, \qquad (2.16)$$

where

$$A \equiv \frac{m}{2r} = \frac{p(1-v^2)}{2[(\bar{x} - v\bar{t})^2 + (1-v^2)(\bar{y}^2 + \bar{z}^2)]^{1/2}}, \qquad (2.17)$$

with $m \equiv p(1-v^2)^{1/2}$. The above metric is not well defined along the light cone when we take the limit $v \to 1$. To overcome this problem, Aichelburg and Sexl introduced the new coordinates t' and x' via the relations,

$$x' - vt' = \bar{x} - v\bar{t}, \quad y' = \bar{y}, \quad z' = \bar{z},$$
$$x' + vt' = \bar{x} + v\bar{t} - 4p\ln[\sqrt{(\bar{x} - \bar{t})^2 + (1-v^2)} - (\bar{x} - \bar{t})], \qquad (2.18)$$

which lead the metric (2.16) to take the form,

$$ds^2 = dt'^2 - dx'^2 - dy'^2 - dz'^2 - 4p\left[\frac{1}{\sqrt{(x' - vt')^2 + \rho^2(1-v^2)}}\right.$$
$$\left. - \frac{1}{\sqrt{(x' - vt')^2 + (1-v^2)}}\right](dt' - dx')^2, \qquad (2.19)$$

where $\rho^2 \equiv y'^2 + z'^2$. Using the limit,

$$\lim_{v \to 1} \left[\frac{1}{\sqrt{(x' - vt')^2 + \rho^2(1 - v^2)}} - \frac{1}{\sqrt{(x' - vt')^2 + (1 - v^2)}} \right]$$
$$= -2ln(\rho)\delta(x' - t'), \tag{2.20}$$

we find that

$$\lim_{v \to 1} ds^2 = dt'^2 - dx'^2 - dy'^2 - dz'^2$$
$$+ 4p\delta(t' - x') \ln(y'^2 + z'^2)(dt' - dx')^2, \tag{2.21}$$

which takes precisely the form of Eq. (2.10) by setting $(\hat{u}, Z, X, Y) = (t' - x', (t' + x')/2, y', z')$.

2.4. Polarization of Gravitational Plane Waves

By means of a coordinate transformation (Hoenselaers and Ernst, 1990), it can be shown that the metric given by Eqs. (2.10) and (2.13) is brought to the form

$$ds^2 = 2e^{-M}dudv - e^{-U}\{e^V \cosh W(dx^2)^2 - 2\sinh w dx^2 dx^3$$
$$+ e^{-V} \cosh W(dx^3)^2\}, \tag{2.22}$$

where M, U, V and W are functions of either the null coordinate u or v. When $W = 0$, Eq. (2.22) reduces to the form for the linearly polarized plane gravitational wave, studied by Rosen (1937).

The spacetime for a plane gravitational wave is described by the metric (2.22) in terms of the coordinates (u, v, x^2, x^3). We first consider the cases in which M, U, V and W are functions of u only. We choose the null tetrad as (Szekeres, 1972; Griffiths, 1991; Wang, 1991f)

$$l^\mu = B\{0, 1, 0, 0\},$$
$$n^\mu = A\{1, 0, 0, 0\}, \tag{2.23}$$
$$m^\mu = \{0, 0, \zeta^2, \zeta^3\},$$
$$\overline{m}^\mu = \{0, 0, \overline{\zeta^2}, \overline{\zeta^3}\},$$

where

$$\zeta^2 \equiv \frac{e^{(U-V)/2}}{\sqrt{2}}\left\{\cosh\frac{W}{2} + i\sinh\frac{W}{2}\right\},$$
$$\zeta^3 \equiv \frac{e^{(U+V)/2}}{\sqrt{2}}\left\{\sinh\frac{W}{2} + i\cosh\frac{W}{2}\right\}, \tag{2.24}$$

and

$$M = \ln(AB). \tag{2.25}$$

From Eq. (1.69), it can be shown that the non-vanishing spin coefficients
are given by

$$\mu = -\frac{1}{2}AU_{,u}, \quad \gamma = \frac{1}{2}A[(\ln B)_{,u} - (1/2)iV_{,u}\sinh W],$$

$$\lambda = \frac{1}{2}A[V_{,u}\cosh W + iW_{,u}], \tag{2.26}$$

where $U_{,u} \equiv \partial U/\partial u$, and so on. Then, from Eqs. (1.77a)–(1.77r) we find
that the non-vanishing Weyl and Ricci scalars are given, respectively, by

$$\Psi_4 = -\Delta\lambda - \lambda(\mu + \overline{\mu} + 3\gamma - \overline{\gamma}), \tag{2.27a}$$

$$\Phi_{22} = -\Delta\mu - \mu(\gamma + \overline{\gamma}) - \lambda\overline{\lambda}. \tag{2.27b}$$

In the vacuum case, we have the Einstein field equation

$$\Phi_{22} = 0, \tag{2.28}$$

where Φ_{22} is given explicitly in terms of the metric coefficients by Eq. (3.8d)
in Chapter 3.

On the other hand, from Eqs. (1.82) and (2.26) we see that in the
present case the null vector n_μ defines a null geodesic congruence, and if B
is chosen to be constant, then the null geodesics are affinely parameterized.
Thus, when M, U, V and W are functions of the null coordinate u only,
the Petrov type N plane gravitational wave represented by Ψ_4 propagates
along the null geodesics (see Fig. 2.1).

The Weyl tensor given by Eq. (1.70) now takes the following form:

$$C_{\mu\nu\lambda\sigma} = -4\{\Psi_4 l_{[\mu}m_{\nu]}l_{[\lambda}m_{\sigma]} + \overline{\Psi}_4 l_{[\mu}\overline{m}_{\nu]}l_{[\lambda}\overline{m}_{\sigma]}\}. \tag{2.29}$$

If we define $m_{\mu\nu}^{(1)}$ and $m_{\mu\nu}^{(2)}$ as (Wang, 1991f)

$$m_{\mu\nu}^{(1)} = 2l_{[\mu}E_{(2)\nu]}, \quad m_{\mu\nu}^{(2)} = 2l_{[\mu}E_{(3)\nu]}, \tag{2.30}$$

where $E_{(2)\mu}$ and $E_{(3)\mu}$ are given by Eq. (1.85), we find that Eq. (2.29) can
be written in the following form:

$$C_{\mu\nu\lambda\sigma} = -\frac{1}{2}\{[m_{\mu\nu}^{(1)}m_{\lambda\sigma}^{(1)} - m_{\mu\nu}^{(2)}m_{\lambda\sigma}^{(2)}](\Psi_4 + \overline{\Psi}_4)$$

$$+ i[m_{\mu\nu}^{(1)}m_{\lambda\sigma}^{(2)} + m_{\mu\nu}^{(2)}m_{\lambda\sigma}^{(1)}](\Psi_4 - \overline{\Psi}_4)\}. \tag{2.31}$$

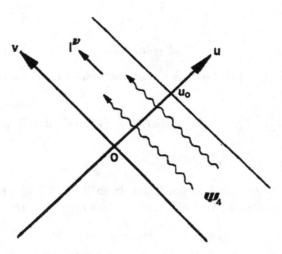

Fig. 2.1.   The spacetime for a plane gravitational wave. For a sandwich wave, the matter can be arranged so that the regions $u < 0$, and $u > u_0$ are flat, and the hypersurface $u = 0$ is the leading wavefront of the wave.

Making a coordinate transformation in the $(E_{(2)\mu}, E_{(3)\mu})$-plane

$$
\begin{aligned}
E_{(2)\mu} &= \cos\varphi_4 E'_{(2)\mu} + \sin\varphi_4 E'_{(3)\mu}, \\
E_{(3)\mu} &= -\sin\varphi_4 E'_{(2)\mu} + \cos\varphi_4 E'_{(3)\mu},
\end{aligned}
\tag{2.32}
$$

we find that

$$
\begin{aligned}
m_{\mu\nu}^{(1)} &= \cos\varphi_4 m_{\mu\nu}'^{(1)} + \sin\varphi_4 m_{\mu\nu}'^{(2)}, \\
m_{\mu\nu}^{(2)} &= -\sin\varphi_4 m_{\mu\nu}'^{(1)} + \cos\varphi_4 m_{\mu\nu}'^{(2)},
\end{aligned}
\tag{2.33}
$$

and

$$
\begin{aligned}
e_{+\mu\nu\lambda\sigma} &= \cos 2\varphi_4 e'_{+\mu\nu\lambda\sigma} + \sin 2\varphi_4 e'_{\times\mu\nu\lambda\sigma}, \\
e_{\times\mu\nu\lambda\sigma} &= -\sin 2\varphi_4 e'_{+\mu\nu\lambda\sigma} + \cos 2\varphi_4 e'_{\times\mu\nu\lambda\sigma},
\end{aligned}
\tag{2.34}
$$

where

$$
\begin{aligned}
e_{+\mu\nu\lambda\sigma} &\equiv m_{\mu\nu}^{(1)} m_{\lambda\sigma}^{(1)} - m_{\mu\nu}^{(2)} m_{\lambda\sigma}^{(2)}, \\
e_{\times\mu\nu\lambda\sigma} &\equiv m_{\mu\nu}^{(1)} m_{\lambda\sigma}^{(2)} + m_{\mu\nu}^{(2)} m_{\lambda\sigma}^{(1)}.
\end{aligned}
\tag{2.35}
$$

Thus, in terms of $E'_{(2)\mu}$ and $E'_{(3)\mu}$, Eq. (2.31) reads

$$
C_{\mu\nu\lambda\sigma} = -\{[\cos 2\varphi_4 \,\mathrm{Re}(\Psi_4) + \sin 2\varphi_4 \,\mathrm{Im}(\Psi_4)]e'_{+\mu\nu\lambda\sigma} \\
+ [\sin 2\varphi_4 \,\mathrm{Re}(\Psi_4) - \cos 2\varphi_4 \,\mathrm{Im}(\Psi_4)]e'_{x\mu\nu\lambda\sigma}\}.
\tag{2.36}
$$

If we choose the angle φ_4 such that

$$\sin 2\varphi_4 \operatorname{Re}(\Psi_4) - \cos 2\varphi_4 \operatorname{Im}(\Psi_4) = 0, \tag{2.37}$$

or equivalently

$$\tan 2\varphi_4 = \frac{\operatorname{Im}(\Psi_4)}{\operatorname{Re}(\Psi_4)}, \tag{2.38}$$

we obtain

$$C_{\mu\nu\lambda\sigma} = -(\Psi_4\overline{\Psi}_4)^{1/2} e'_{+\mu\nu\lambda\sigma}. \tag{2.39}$$

Defining the null vector $\hat{l}^\mu$ as

$$\hat{l}^\mu = (\Psi_4\overline{\Psi}_4)^{1/4} l^\mu, \tag{2.40}$$

Eq. (2.39) becomes

$$C_{\mu\nu\lambda\sigma} = -\hat{e}_{+\mu\nu\lambda\sigma}. \tag{2.41}$$

Hence, following Eq. (2.6) we find that for a time-like geodesic congruence the geodesic deviation is given by

$$\begin{aligned}
\frac{D^2\eta^\mu}{D\tau^2} &= (\hat{l}_\lambda u^\lambda)^2 [E'^\mu_{(2)} E'^\nu_{(2)} - E'^\mu_{(3)} E'^\nu_{(3)}]\eta_\nu \\
&= (\Psi_4\overline{\Psi}_4)^{1/2}(l_\lambda u^\lambda)^2 [E'^\mu_{(2)} E'^\nu_{(2)} - E'^\mu_{(3)} E'^\nu_{(3)}]\eta_\nu.
\end{aligned} \tag{2.42}$$

Equation (2.42) shows that the relative accelerations of geodesics are proportional to $(\Psi_4\overline{\Psi}_4)^{1/2}$, which does not relate to any observer. Thus, $(\Psi_4\overline{\Psi}_4)^{1/2}$ represents the absolute amplitude of the relative accelerations of neighboring test particles. The angle φ_4 is the polarization angle of the plane gravitational wave with respect to the basis, $(E_{(2)\mu}, E_{(3)\mu})$.

For the spacetimes described by the metric (2.22), it is easy to show that

$$E^\mu_{(2);\nu} l^\nu = 0 = E^\mu_{(3);\nu} l^\nu. \tag{2.43}$$

That is, the basis $(E_{(2)\mu}, E_{(3)\mu})$ is parallelly transported along the null geodesics defined by n_μ. Since Ψ_4 is a function of u only, we have

$$\varphi_{4,v} = \frac{1}{2}\left(\tan^{-1}\frac{\operatorname{Im}\Psi_4}{\operatorname{Re}\Psi_4}\right)_{,v} = 0. \tag{2.44}$$

It follows that the polarization angle φ_4 is constant along the path that the plane gravitational wave follows, which is consistent with the definition of a plane gravitational wave given in Section 2.2.

In a similar fashion, it can be shown that the non-vanishing Weyl and Ricci scalars are Ψ_0 and Φ_{00}, when M, U, V and W are functions of the null coordinate v only, and that the geodesic deviation for a time-like geodesic congruence is given by

$$\frac{D^2\eta^\mu}{D\tau^2} = (\Psi_0\overline{\Psi}_0)^{1/2}(n_\lambda u^\lambda)^2[E'^\mu_{(2)}E''^\nu_{(2)} - E'^\mu_{(3)}E''^\nu_{(3)}]\eta_\nu, \qquad (2.45)$$

but now with the angle φ_0 being defined as

$$\tan 2\varphi_0 = -\frac{\mathrm{Im}(\Psi_0)}{\mathrm{Re}(\Psi_0)}. \qquad (2.46)$$

Similar to Eq. (2.43), we have

$$E^\mu_{(2);\nu}n^\nu = 0 = E^\mu_{(3);\nu}n^\nu. \qquad (2.47)$$

Thus, in the latter case the metric (2.22) describes a plane gravitational wave represented by Ψ_0 with the polarization angle φ_0, which moves along the null geodesic congruence defined by l_μ. The angle φ_0 is constant along the path, along which the Ψ_0-wave propagates.

When the angle φ_0 (φ_4) is constant everywhere, we say the corresponding plane gravitational wave is constantly polarized.

2.5. Singularities of Gravitational Plane Wave Spacetimes and Memory Effects

The memory effects of GWs have been attracted lots of attention (see, for example, Favata, 2010; Bieri, Garfinkle and Yunes, 2017; Zhang, Duval, Gibbons and Horvathy, 2017; and references therein), especially after the recent observations of several GWs emitted from remote binary systems of either black holes or neutron stars (Abbott *et al.*, 2019). Such effects might be possibly detected by LISA (Favata, 2010) or even by current generation of detectors, such as the advanced LIGO and Virgo (Lasky *et al.*, 2016). Recently, such investigations gained new momenta due to the close relations between asymptotically symmetric theorems of soft gravitons and GW memory effects (Hawking, Perry and Strominger, 2016, 2017; Strominger, 2017).

The characteristic feature of these effects is the permanent displacement of a test particle after a burst of a GW passes (Braginsky and Grishchuk, 1985; Christodoulou, 1991; Blanchet and Damour, 1992; Thorne, 1992;

Harte, 2013). In addition, the passage of the GW affects not only the position of the test particle, but also its velocity. In fact, the change of the velocity of the particle is also permanent (Souriau, 1973; Braginsky and Thorne, 1987; Bondi, 1957; Bondi and Pirani, 1989; Grishchuk and Polnarev, 1989; Zhang, Duval, Gibbons and Horvathy, 2018).

When far from the sources, the emitted GWs can be well approximated by plane GWs. The spacetimes for plane GWs can be cast in various forms, depending on the choice of the coordinates and gauge-fixing, as showed in Sections 2.2 and 2.4. The form of Eq. (2.22) was originally due to Baldwin, Jeffery and Rosen (BJR) (Baldwin and Jeffery, 1926; Rosen, 1937). Despite its several attractive features, the system of the BJR coordinates is often singular within a finite width of a wave, and when studying the asymptotic behavior of the spacetime, extensions beyond this singular surface are needed.

In this section, we point out that there exist actually two kinds of singularities in plane gravitational wave spacetimes, one represents coordinate singularities, which can be removed by proper coordinate transformations, and the other represents really spacetime singularities, and physical quantities, such as distortions of test particles, become infinitely large when such singularities are approaching (Wang *et al.*, 2018). Therefore, in the latter these singularities already represent the boundaries of the spacetimes and extensions beyond them are not only impossible but also not needed. Since gravitational memory effects and soft graviton theorems are closely related to the asymptotical behaviors of plane GW spacetimes, in the latter the spacetimes cannot be used to study such properties.

In GR, there are powerful Hawking–Penrose theorems (Hawking and Ellis, 1973) from which one can see that spacetimes with "physically reasonable" conditions are singular. However, the theorems did not tell the nature of the singularities, and Ellis and Schmidt (1977) classified them into two different kinds: *spacetime curvature singularities* and *coordinate singularities*. The former is real and cannot be removed by any coordinate transformations of the kind,

$$x^\mu \to x'^\mu = \zeta^\mu(x^\nu), \quad \mu, \nu = 0, 1, 2, 3, \tag{2.48}$$

while the latter is coordinate dependent, and can be removed by proper coordinate transformations. One typical example is the coordinate singularity of the Schwarzschild solution in the spherical coordinates at the Schwarzschild radius $r = 2MG/c^2$.

Spacetime curvature singularities are further divided into two sub-classes: *scalar curvature singularities* and *non-scalar curvature singularities*. If any of the 14 independent scalars (Campbell and Wainwright, 1977), constructed from the four-dimensional Riemann tensor $R^{\sigma}_{\mu\nu\lambda}$ and its derivatives, is singular, then the spacetime is said singular, and the corresponding singularity is a scalar one. If none of these scalars is singular, spacetimes can be still singular. In particular, tidal forces and/or distortions (which are the double integrals of the tidal forces), experienced by an observer, may become infinitely large (Ori, 2000; Nolan, 2000; Hirschmann, Wang and Wu, 2004; Sharma, Tziolas, Wang and Wu, 2011). This kind of singularities is usually referred to as non-scalar curvature singularities.

In the spacetimes of plane GWs, all the 14 independent scalars vanish identically (Stephani *et al.*, 2009), so in such spacetimes the singularities can be either non-scalar (but real spacetime) singularities or coordinate singularities. By studying tidal forces and distortions of freely falling observers, Wang *et al.* (2018) recently showed that the spacetimes are not singular only in some particular cases. To show this, for the sake of simplicity, we can consider only the diagonal case, in which $W = 0$ in the metric Eq. (2.22). Then, the only non-vanishing component is R_{uu}, given by

$$R_{uu} = U'' - \frac{1}{2}(U'^2 + V'^2). \qquad (2.49)$$

Thus, in the vacuum spacetimes, we have $R_{uu} = 0$, which yields

$$\chi'' + \omega^2\chi = 0, \qquad (2.50)$$

where

$$\chi \equiv e^{-U/2}, \quad \omega \equiv \frac{1}{2}V'. \qquad (2.51)$$

Then, from Eq. (2.50) we can see that, for any given initial value, $\chi_0 > 0$, there always exists a moment, say, $u = u_s$ at which χ vanishes,

$$\chi(u_s) = 0 \quad \text{or} \quad U(u_s) = +\infty, \qquad (2.52)$$

that is, a singularity of the metric (2.22) appears at $u = u_s$, which is surely not a scalar singularity, as mentioned above, all the 14 independent scalars made of the Riemann tensor now vanish identically. Does this mean that the singularity must be a coordinate one? The answer is not always affirmative. This is because spacetimes can still have non-scalar singularities, as mentioned above. The non-scalar spacetime singularities can be indicated by, for example, the divergence of distortions of a freely falling observer, which are the twice integrations of the tidal force with respect to the proper time of the observer (Ori, 2000; Nolan, 2000; Hirschmann, Wang and Wu, 2004; Sharma, Tziolas, Wang and Wu, 2011).

To show this, one can first write $\chi(u)$ near the focusing point $u = u_s$ as

$$\chi(u) \equiv e^{-U(u)/2} = (u - u_s)^\alpha \hat{\chi}(u), \tag{2.53}$$

where $\alpha > 0$, and $\hat{\chi}(u_s) \neq 0$. The function $\hat{\chi}(u)$ in general takes the form,

$$\hat{\chi}(u) = \sum_{n=0}^{\infty} \chi_n (u - u_s)^n, \tag{2.54}$$

with $\chi_0 \neq 0$. Then, one can consider the tetrad, $e^\mu_{(a)}$ $(a = 0, 1, 2, 3)$, defined by

$$e^\mu_{(0)} = \gamma_0 \delta^\mu_u + \frac{1}{2\gamma_0} \delta^\mu_v, \quad e^\mu_{(1)} = \gamma_0 \delta^\mu_u - \frac{1}{2\gamma_0} \delta^\mu_v,$$
$$e^\mu_{(2)} = e^{\frac{U-V}{2}} \delta^\mu_y, \quad e^\mu_{(3)} = e^{\frac{U+V}{2}} \delta^\mu_z, \tag{2.55}$$

which satisfies the relations,

$$e^\mu_{(\alpha)} e^\nu_{(\beta)} g_{\mu\nu} = \eta_{\alpha\beta}, \quad e^\mu_{(\alpha);\nu} e^\nu_{(0)} = 0, \tag{2.56}$$

that is, they are unit orthogonal vectors and parallelly transported along the time-like geodesics, defined by $e^\mu_{(0)} \equiv dx^\mu/d\lambda$, where λ denotes the proper time of the time-like geodesics, so that they form a freely falling frame (Ori, 2000; Nolan, 2000; Hirschmann, Wang and Wu, 2004; Sharma, Tziolas, Wang and Wu, 2011). Projecting the Riemann tensor onto this frame, one obtains some non-zero components of the Riemann tensor $R_{(a)(b)(c)(d)}$. If one integrates them twice along the time-like geodesics, which gives the distortions, one finds that such integral always diverges as $u \to u_s$, except for the cases (Wang *et al.*, 2018),

$$\text{(i)} \ \alpha = \frac{1}{2} \quad \text{or} \quad \text{(ii)} \ \alpha = 1. \tag{2.57}$$

Therefore, *all the plane GW spacetimes are singular physically at the focused point $u = u_s$, exceptions are only the ones with $\alpha = 1/2$ or 1.* As a result, all the plane GW spacetimes cannot be used to study memory effects and soft graviton theorems, except these two particular cases, as only these space-times can be possibly extended to null infinity, whereby memory effects and soft graviton theorems can be studied.

The above results, although very simple, may have profound implications on the studies of gravitational memory effects (Favata, 2010; Bieri, Garfinkle and Yunes, 2017) and "soft-graviton" theorems (Hawking, Perry and Strominger, 2016, 2017; Strominger, 2017), as both of them are concerned with the asymptotic behavior of the spacetimes at infinities. But when the spacetimes become singular at the finite focusing point, such infinities do not exist. Therefore, only the non-singular spacetimes are relevant to them.

Chapter 3

Colliding Plane Gravitational Waves

In this chapter, we provide a general description for the spacetimes of colliding plane gravitational waves in vacuum or coupled to matter fields. Specifically, in Section 3.1 the spacetimes for two colliding plane gravitational waves are discussed, and the connection coefficients are explicitly given in terms of the metric coefficients M, U, V and W, and their derivatives. Then, in Section 3.2 the spin coefficients, Weyl and Ricci scalars, and Bianchi identities are given in terms of distributions, by assuming that across the wavefronts of the two incoming gravitational waves, the functions M, U, V and W are only continuous, that is, C^0. The different coordinate systems are discussed in Section 3.3, and following it, in Section 3.4, the polarization of an interacting gravitational wave is defined with respect to a parallelly transported frame along the wave path, and then the gravitational analogy of the well-known Faraday rotation in electrodynamics is studied in detail. In Section 3.5, the nature of spacetime singularities formed due to the mutual focus of the two colliding plane gravitational waves is investigated, and finally in Section 3.6 several methods for generating new solutions of the Einstein field equations are reviewed.

3.1. Spacetimes for Colliding Plane Gravitational Waves

In Sections 2.2 and 2.3, we discussed the spacetimes for plane gravitational waves. It was shown that when the functions M, U, V and W depend only on the null coordinate u, the metric (2.22) describes a plane gravitational wave moving along $u = $ constant hypersurfaces. While when they depend on the null coordinate v, the metric describes a plane gravitational wave moving along $v = $ constant hypersurfaces. By properly choosing the two null coordinates, the collision of two such plane gravitational waves in the Minkowski background can be always considered as head-on collision. Then, the spacetimes can be arranged as follows.

For $u < 0$ and $v > 0$ (see Fig. 3.1), we assume the metric to be that of a plane wave described by the metric (2.22) with M, U, V and W being functions of v only. Without loss of generality, we can always choose the coordinate v so that the hypersurface $v = 0$ represents the wavefront. For a sandwich wave, we may further assume that the spacetime is flat (correspondingly, M, U, V and W are constants) in the region $v > v_0$. In the region $v < 0$ and $u > 0$, on the other hand, the metric is again assumed to be that of a plane gravitational wave, but with M, U, V and W being functions of u only, and the $u = 0$ hypersurface is the wavefront. For a sandwich wave, it is further assumed that the region $u > u_0$ is flat. Assuming that the collision occurs in a flat background, then in the region $u, v < 0$ the spacetime is Minkowski, since in this region the two coming waves have not arrived, yet. In the region $u > 0$ and $v > 0$ where the two plane waves interact, the metric coefficients M, U, V and W become functions of both u and v. Hence, a characteristic initial value problem has been set up with the data posed on a pair of null hypersurfaces $u = 0$ and $v = 0$ intersecting in a space-like two-dimensional surface $u = 0 = v$. In principle, these initial conditions uniquely determine the geometry in the wave interacting region $u, v > 0$ (Szekeres, 1972; Xanthopoulos, 1986a; Hauser and Ernst, 1989a, 1989b, 1990). For the sake of convenience, the above various regions are numbered as follows (see Fig. 3.1):

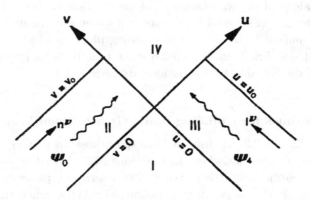

Fig. 3.1.   The projection of a spacetime for colliding plane waves onto the (u, v)-plane.

Region I $(u, v < 0)$: This is the region in which the two colliding plane waves have not arrived, yet, so the spacetime in this region is flat, and the functions M, U, V and W are constants.

Region II $(u < 0, v > 0)$: This is the region in which a plane gravitational wave propagates toward the right-hand side with the $v = 0$ hypersurface as its wavefront. The functions M, U, V and W depend on v only.

Region III $(u > 0, v < 0)$: In this region, an opposite moving plane gravitational wave is incident, and the $u = 0$ hypersurface is its wavefront. In this region M, U, V and W are functions of u only.

Region IV $(u, v > 0)$: In this region two incoming waves interact, and M, U, V and W are functions of both u and v.

In addition to these four regions, there are also two null hypersurfaces, Σ_u and Σ_v, defined as $\Sigma_u \equiv \{x^\mu : u = 0\}$ and $\Sigma_v \equiv \{x^\mu : v = 0\}$, respectively. At the space-like two-dimensional surface $u = 0 = v$, the two plane waves collide. Across the hypersurfaces $u = 0$ and $v = 0$, the Riemann curvature tensor, in general, suffers a shock and or an impulsive type of discontinuity. For the collision of two sandwich waves, Regions II and III reduce, respectively, to $u < 0$ and $0 < v < v_0$, and $0 < u < u_0$ and $v < 0$.

Corresponding to the metric (2.22), the non-vanishing connection coefficients $\Gamma^\mu_{\alpha\beta}$ defined by Eq. (1.8) are given by

$$\Gamma^u_{uu} = -M_{,u}, \quad \Gamma^u_{22} = \frac{1}{2}e^M(e^{V-U}\cosh W)_{,v},$$

$$\Gamma^u_{23} = -\frac{1}{2}e^M(e^{-U}\sinh W)_{,v}, \quad \Gamma^u_{33} = \frac{1}{2}e^M(e^{-V-U}\cosh W)_{,v},$$

$$\Gamma^v_{vv} = -M_{,v}, \quad \Gamma^v_{22} = \frac{1}{2}e^M(e^{V-U}\cosh W)_{,u},$$

$$\Gamma^v_{23} = -\frac{1}{2}e^M(e^{-U}\sinh W)_{,u}, \quad \Gamma^v_{33} = \frac{1}{2}e^M(e^{V-U}\cosh W)_{,u},$$

$$\Gamma^2_{2u} = \frac{1}{2}(\cosh^2 W V_{,u} - U_{,u}), \quad \Gamma^2_{2v} = \frac{1}{2}(\cosh^2 W V_{,v} - U_{,v}),$$

$$\Gamma^3_{3u} = -\frac{1}{2}(\cosh^2 W V_{,u} + U_{,u}), \quad \Gamma^3_{3v} = -\frac{1}{2}(\cosh^2 W V_{,v} + U_{,v}),$$

$$\Gamma^2_{3u} = -\frac{1}{2}e^{-V}(\sinh W \cosh W V_{,u} + W_{,u}),$$

$$\Gamma^2_{3v} = -\frac{1}{2}e^{-V}(\sinh W \cosh W V,_v + W,_v),$$

$$\Gamma^3_{2u} = \frac{1}{2}e^{V}(\sinh W \cosh W V,_u - W,_u),$$

$$\Gamma^3_{2v} = \frac{1}{2}e^{V}[\sinh W \cosh W V,_v - W,_v). \tag{3.1}$$

When $W = 0$, the two incoming gravitational waves have fixed polarization directions, and are said to be collinearly (linearly) polarized. Otherwise, they are said to be non-collinearly (nonlinearly) polarized.

3.2. The Basic Differential Equations for Colliding Plane Gravitational Waves

Assuming that the metric coefficients appearing in Eq. (2.22) are functions of u and v, we find that, corresponding to the choice of the null tetrad (2.23), the spin coefficients defined by Eq. (1.69) are given by

$$\rho = \frac{1}{2}BU,_v, \quad \mu = -\frac{1}{2}AU,_u, \quad \varepsilon = -\frac{1}{4}B[2(\ln A),_v + iV,_v \sinh W],$$

$$\gamma = \frac{1}{4}A[2(\ln B),_u - iV,_u \sinh W], \quad \sigma = -\frac{1}{2}B(V,_v \cosh W - iW,_v), \tag{3.2}$$

$$\lambda = \frac{1}{2}A(V,_u \cosh W + iW,_u),$$

and

$$\kappa = \nu = \tau = \pi = \alpha = \beta = 0. \tag{3.3}$$

The vanishing of the spin coefficients κ and ν implies [see Eq. (1.82)] that the null vectors n^μ and l^μ defined by Eq. (2.23) still define null geodesic congruences even in the interacting region (Region IV), as in the single plane gravitational wave case. Moreover, if the function A is chosen to be constant, then the null geodesics defined by l^μ are affinely parameterized, while when the function B is chosen to be constant, the null geodesics defined by n^μ are affinely parameterized.

The plane gravitational waves represented by Ψ_0 and Ψ_4 propagate along the null geodesic congruences defined, respectively, by n^μ and l^μ. Then, Eqs. (3.2), (1.90) and (1.91) show that after the collision the two null geodesic congruences are no longer shear-free, although their rotations are still zero. Therefore, after the collision, the two plane gravitational waves will in general pick up shear. Substituting Eq. (3.3) into Eqs. (1.77a)–(1.77r), we find

$$D\rho = (\rho^2 + \sigma\overline{\sigma}) + \rho(\varepsilon + \overline{\varepsilon}) + \Phi_{00}, \tag{3.4a}$$

$$D\sigma = \sigma(2\rho + 3\varepsilon - \overline{\varepsilon}) + \Psi_0, \tag{3.4b}$$

$$D\gamma - \Delta\varepsilon = -\gamma(\varepsilon + \overline{\varepsilon}) - \varepsilon(\gamma + \overline{\gamma}) + \Psi_2 + \Phi_{11} - \Lambda, \tag{3.4c}$$

$$D\lambda = \rho\lambda + \overline{\sigma}\mu - \lambda(3\varepsilon - \overline{\varepsilon}) + \Phi_{20}, \tag{3.4d}$$

$$D\mu = \rho\mu + \sigma\lambda - \mu(\varepsilon + \overline{\varepsilon}) + \Psi_2 + 2\Lambda, \tag{3.4e}$$

$$\Delta\lambda = -\lambda(2\mu + 3\gamma - \overline{\gamma}) - \Psi_4, \tag{3.4f}$$

$$0 = \mu\rho - \lambda\sigma - \Psi_2 + \Phi_{11} + \Lambda, \tag{3.4g}$$

$$-\Delta\mu = \mu^2 + \lambda\overline{\lambda} + \mu(\gamma + \overline{\gamma}) + \Phi_{22}, \tag{3.4h}$$

$$-\Delta\sigma = \mu\sigma + \overline{\lambda}\rho - \sigma(3\gamma - \overline{\gamma}) + \Phi_{02}, \tag{3.4i}$$

$$\Delta\rho = -(\rho\mu + \sigma\lambda) + \rho(\gamma + \overline{\gamma}) - \Psi_2 - 2\Lambda, \tag{3.4j}$$

while from Eqs. (1.71)–(1.72) we find that

$$\Psi_1 = \Psi_3 = \Phi_{01} = \Phi_{10} = \Phi_{12} = \Phi_{21} = 0. \tag{3.5}$$

From Eqs. (3.4a)–(3.4j), we obtain

$$\Psi_0 = D\sigma - \sigma(2\rho + 3\varepsilon - \overline{\varepsilon}), \tag{3.6a}$$

$$\Psi_2 = \frac{1}{3}[D\gamma - \Delta(\varepsilon + \rho) + (\rho + \varepsilon)(\gamma + \overline{\gamma}) + \gamma(\varepsilon + \overline{\varepsilon}) - 2\lambda\sigma], \tag{3.6b}$$

$$\Psi_4 = -\Delta\lambda - \lambda(2\mu + 3\gamma - \overline{\gamma}), \tag{3.6c}$$

$$\Phi_{00} = D\rho - (\rho^2 + \sigma\overline{\sigma}) - \rho(\varepsilon + \overline{\varepsilon}), \tag{3.6d}$$

$$\Phi_{11} = \frac{1}{2}[D\gamma - \Delta\varepsilon + \gamma(\varepsilon + \overline{\varepsilon}) + \varepsilon(\gamma + \overline{\gamma}) + \lambda\sigma - \rho\mu], \tag{3.6e}$$

$$\Phi_{02} = -\Delta\sigma + \sigma(3\gamma - \overline{\gamma}) - \mu\sigma - \overline{\lambda}\rho, \tag{3.6f}$$

$$\Phi_{22} = -\Delta\mu - \mu(\gamma + \overline{\gamma}) - \mu^2 - \lambda\overline{\lambda}, \tag{3.6g}$$

$$\Lambda = \frac{1}{6}[\Delta(\varepsilon - 2\rho) - D\gamma - \gamma(\varepsilon + \overline{\varepsilon})$$
$$+ (2\rho - \varepsilon)(\gamma + \overline{\gamma}) - \lambda\sigma - 3\mu\rho]. \tag{3.6h}$$

Substituting Eq. (3.2) into Eqs. (3.6a)–(3.6h), we find

$$\Psi_0 = -\frac{1}{2}B^2\{V_{,vv}\cosh W + (M_{,v} - U_{,v})V_{,v}\cosh W + 2\sinh W V_{,v}W_{,v}$$
$$- i[W_{,vv} + (M_{,v} - U_{,v})W_{,v} - \sinh W \cosh W V_{,v}{}^2]\}, \qquad (3.7a)$$

$$\Psi_2 = \frac{1}{6}AB\left\{M_{,uv} - U_{,uv} + W_{,u}W_{,v} + \cosh^2 W V_{,u}V_{,v}\right.$$
$$\left. + i\frac{3}{2}\cosh W (V_{,u}W_{,v} - V_{,v}W_{,u})\right\}, \qquad (3.7b)$$

$$\Psi_4 = -\frac{1}{2}A^2\{V_{,uu}\cosh W + (M_{,u} - U_{,u})V_{,u}\cosh W + 2\sinh W V_{,u}W_{,u}$$
$$+ i[W_{,uu} + (M_{,u} - U_{,u})W_{,u} - \sinh W \cosh W V_{,u}^2]\}, \qquad (3.7c)$$

and

$$\Phi_{00} = \frac{1}{4}B^2\{2U_{,vv} - U_{,v}{}^2 + 2U_{,v}M_{,v} - W_{,v}{}^2 - \cosh^2 W V_{,v}{}^2\}, \qquad (3.8a)$$

$$\Phi_{11} = \frac{1}{8}AB\{2M_{,uv} + U_{,u}U_{,v} - W_{,u}W_{,v} - \cosh^2 W V_{,u}V_{,v}\}, \qquad (3.8b)$$

$$\Phi_{02} = \frac{1}{4}AB\{2\cosh W V_{,uv} - \cosh W (U_{,u}V_{,v} + V_{,u}U_{,v})$$
$$+ 2\sinh W (V_{,u}W_{,v} + V_{,v}W_{,u}) - i[2W_{,uv} - (U_{,u}W_{,v} + W_{,u}U_{,v})$$
$$- 2\sinh W \cosh W V_{,u}V_{,v}]\}, \qquad (3.8c)$$

$$\Phi_{22} = \frac{1}{4}A^2\{2U_{,uu} - U_{,u}{}^2 + 2U_{,u}M_{,u} - W_{,u}{}^2 - \cosh^2 W V_{,u}{}^2\}, \qquad (3.8d)$$

$$\Lambda = -\frac{1}{24}AB\{2M_{,uv} + 4U_{,uv} - 3U_{,u}U_{,v} - W_{,u}W_{,v}$$
$$- \cosh^2 W V_{,u}V_{,v}\}. \qquad (3.8e)$$

On the other hand, from Eqs. (1.78a)–(1.78k), (3.3) and (3.5) we find that the Bianchi identities now read

$$\Delta\Psi_0 + D\Phi_{02} = (4\gamma - \mu)\Psi_0 + 3\sigma\Psi_2 + (2\varepsilon - 2\bar{\varepsilon} + \rho)\Phi_{02}$$
$$+ 2\sigma\Phi_{11} - \bar{\lambda}\Phi_{00}, \qquad (3.9a)$$

$$3D\Psi_2 - 2D\Phi_{11} + \Delta\Phi_{00} = -3\lambda\Psi_0 + 9\rho\Psi_2 + (\mu + 2\gamma + 2\overline{\gamma})\Phi_{00}$$
$$- 2\rho\Phi_{11} - 2\sigma\Phi_{20} + \overline{\sigma}\Phi_{02}, \tag{3.9b}$$

$$3\Delta\Psi_2 - 2\Delta\Phi_{11} + D\Phi_{22} = 3\sigma\Psi_4 - 9\mu\Psi_2 - (\rho + 2\varepsilon + 2\overline{\varepsilon})\Phi_{22}$$
$$+ 2\mu\Phi_{11} + 2\lambda\Phi_{02} - \overline{\lambda}\Phi_{20}, \tag{3.9c}$$

$$D\Psi_4 + \Delta\Phi_{20} = (\rho - 4\varepsilon)\Psi_4 - 3\lambda\Psi_2 - (\mu + 2\gamma - 2\overline{\gamma})\Phi_{20}$$
$$- 2\lambda\Phi_{11} + \overline{\sigma}\Phi_{22}, \tag{3.9d}$$

and

$$D\Phi_{11} + \Delta\Phi_{00} + 3D\Lambda = 2(\gamma + \overline{\gamma} - \mu)\Phi_{00}$$
$$+ 4\rho\Phi_{11} + \overline{\sigma}\Phi_{02} + \sigma\Phi_{20}, \tag{3.10a}$$

$$\Delta\Phi_{11} + D\Phi_{22} + 3\Delta\Lambda = 2(\rho - \varepsilon - \overline{\varepsilon})\Phi_{22}$$
$$- 4\mu\Phi_{11} - \overline{\lambda}\Phi_{20} - \lambda\Phi_{02}. \tag{3.10b}$$

Equations (3.7a)–(3.10b) are the basic differential equations for colliding plane gravitational waves. To solve them for a given source usually follows two different paths. One is to pose the initial data on the two half hypersurfaces $u = 0, v \geq 0$ and $u \geq 0, v = 0$, and then solve the corresponding initial value problem (Szekeres, 1972; Yurtsever, 1988a, 1988b, 1989; Griffiths, 1991). However, this method is not practical in obtaining exact solutions, especially in the non-collinear case ($W \neq 0$). Another method which was pioneered by Khan and Penrose (1971) is essentially to work backward in time. That is, we first find a solution that is valid in the interaction region (Region IV), and then extend it back to the pre-collision regions (Regions I–III) by means of the following substitutions:

$$u \to uH(u), \quad v \to vH(v), \tag{3.11}$$

in the metric coefficients, where $H(x)$ denotes the Heaviside function,

$$H(x) = \begin{cases} 1, & x > 0, \\ 0, & x < 0. \end{cases} \tag{3.12}$$

However, such an extension can guarantee the metric coefficients only be C^0 across the null hypersurfaces $u = 0$ and $v = 0$. To be physically acceptable, we must impose some conditions on these coefficients.

We assume that all the metric coefficients are at least C^3 in the interaction region (Region IV), and C^0 across the hypersurfaces $u = 0$ or $v = 0$. Then, as shown in Section 1.9, the Einstein field equations (1.77a)–(1.77r) as well as the Bianchi identities (1.78a)–(1.78k) hold in the sense of distributions. For any given function f, which is at least C^3 in Region IV and C^0 across the hypersurfaces $u = 0$ or $v = 0$, it can be written in the form (Wang, 1991f, 1992a),[1]

$$f(uH(u), vH(v)) = f(u,v)H(u)H(v) + f(0,v)[1 - H(u)]H(v)$$

$$+ f(u,0)[(1 - H(v)]H(u), \tag{3.13}$$

where

$$f(0,v) = \lim_{u \to 0^+} f(u,v), \quad f(u,0) = \lim_{v \to 0^+} f(u,v). \tag{3.14}$$

Thus, for any given test function $F(u,v)$, we have

$$\langle \partial_u f(uH(u), vH(v)), F(u,v) \rangle$$

$$= \langle \partial_u \{ f(u,v)H(u)H(v) + f(0,v)[1 - H(u)]H(v)$$

$$+ f(u,0)[1 - H(v)]H(u) \}, F(u,v) \rangle$$

$$= \langle f_{,u}(u,v)H(u)H(v) + f_{,u}(u,0)[1 - H(v)]H(u), F(u,v) \rangle$$

$$+ \int_\Omega F(u,v)f(u,v)H(v)[\partial_u H(u)]dV$$

$$+ \int_\Omega F(u,v)f(0,v)H(v)[\partial_u[1 - H(u)]]dV$$

$$+ \int_\Omega F(u,v)f(u,0)(1 - H(v)[\partial_u H(u)]dV$$

$$= \langle f_{,u}(u,v)H(u)H(v) + f_{,u}(u,0)[1 - H(v)]H(u), F(u,v) \rangle$$

$$- \int_{\Omega+u} \partial_u[F(u,v)f(u,v)H(v)]dV - \int_{\Omega-u} \partial_u[F(u,v)f(0,v)H(v)]dV$$

$$- \int_{\Omega+u} \partial_u[F(u,v)f(u,0)[1 - H(v)]]dV$$

$$= \langle f_{,u}(u,v)H(u)H(v) + f_{,u}(u,0)[1 - H(v)]H(u), F(u,v) \rangle$$

---

[1] A similar expression was also applied to domain walls (Wang, 1991c, 1992e, 1993; Schmidt and Wang, 1993; Letelier and Wang, 1994) and brane worlds in string/M-theory (Gong, Wang and Wu, 2008; Tziolas and Wang, 2008; Wang and Santos, 2008; Wu *et al.*, 2008; Wu, Gong and Wang, 2009; Tziolas, Wang and Wu, 20009; Devin *et al.*, 2009; Wang and Santos, 2010; Wang, 2010; Sharma *et al.*, 2011), which turns out to be very useful in the studies of singular surfaces.

$$-\int_{\partial\Omega^{+u}} F(0,v)H(v)[f^+(0,v) - f^-(0,v)]dS_u$$

$$-\int_{\partial\Omega^{+u}} F(0,v)f(0,0)[1 - H(v)]dS_u, \qquad (3.15)$$

where $F(u,v)$ is at least a C^1-function with compact support on Ω. Since f is C^0 across the hypersurface $u = 0$, we have $f^+(0,v) = f^-(0,v)$, and Eq. (3.15) becomes

$$\langle \partial_u f(uH(u), vH(v)), F(u,v) \rangle$$
$$= \langle f_{,u}(u,v)H(u)H(v) + f_{,u}(u,0)[1 - H(v)]H(u), F(u,v) \rangle$$
$$- \int_{\partial\Omega^{+u}} F(0,v)f(0,0)[1 - H(v)]dS_u. \qquad (3.16)$$

If we further assume that

$$f(0,0) = 0, \qquad (3.17)$$

then Eq. (3.16) can be written as

$$\langle \partial_u f(uH(u), vH(v)), F(u,v) \rangle$$
$$= \langle f_{,u}(u,v)H(u)H(v) + f_{,u}(u,0)[1 - H(v)]H(u), F(u,v) \rangle, \qquad (3.18)$$

or equivalently

$$f_{,u}(uH(u), vH(v)) = f_{,u}(u,v)H(u)H(v) + f_{,u}(u,0)[1 - H(v)]H(u)$$
$$= f_{,u}(u, vH(v))H(u). \qquad (3.19)$$

In a similar fashion, it can be shown that

$$f_{,v}(uH(u), vH(v)) = f_{,v}(uH(u), v)H(v),$$
$$f_{,vu}(uH(u), vH(v)) = f_{,uv}(u,v)H(v)H(u),$$
$$f_{,uu}(uH(u), vH(v)) = f_{,uu}(u,v)H(v)H(u) + f_{,uu}(u,0)[1 - H(v)]H(u)$$
$$+ \delta(u)f_{,u}(0,v)H(v) + \delta(u)f_{,u}(0,0)[1 - H(v)]$$
$$= f_{,uu}(u, vH(v))H(u) + \delta(u)f_{,u}(0, vH(v)),$$
$$f_{,vv}(uH(u), vH(v)) = f_{,vv}(u,v)H(v)H(u) + f_{,vv}(0,v)[1 - H(u)]H(v)$$
$$+ \delta(v)f_{,v}(u,0)H(u) + \delta(v)f_{,v}(0,0)[(1 - H(u)]$$
$$= f_{,vv}(uH(u), v)H(v) + \delta(v)f_{,v}(uH(u), 0), \qquad (3.20)$$

where

$$f_{,u}(0, vH(v)) = \lim_{u \to 0^+} f_{,u}(u, vH(v)),$$

$$f_{,v}(uH(u), 0) = \lim_{v \to 0^+} f_{,v}(uH(u), v). \tag{3.21}$$

Therefore, if we assume that

$$M(0,0) = U(0,0) = V(0,0) = W(0,0) = 0, \tag{3.22}$$

then the non-vanishing spin coefficients given by Eq. (3.2) can be written as

$$\rho = \frac{1}{2} B(uH(u), v) U_{,v}(uH(u), v) H(v),$$

$$\mu = -\frac{1}{2} A(u, vH(v)) U_{,u}(u, vH(v)) H(u),$$

$$\varepsilon = -\frac{1}{2} B(uH(u), v) \left\{ [\ln A(uH(u), v)]_{,v} \right.$$

$$\left. + \frac{i}{2} V_{,v}(uH(u), v) \sinh W(uH(u), v) \right\} H(v),$$

$$\gamma = \frac{1}{2} A(u, vH(v)) \left\{ [\ln B(u, vH(v))]_{,u} \right. \tag{3.23}$$

$$\left. - \frac{i}{2} V_{,u}(u, vH(v)) \sinh W(u, vH(v)) \right\} H(u),$$

$$\sigma = -\frac{1}{2} B(uH(u), v) \{ V_{,v}(uH(u), v) \cosh W(uH(u), v)$$

$$- iW_{,v}(uH(u), v) \} H(v),$$

$$\lambda = \frac{1}{2} A(u, vH(v)) \{ V_{,u}(u, vH(v)) \cosh W(u, vH(v))$$

$$+ iW_{,u}(u, vH(v)) \} H(u).$$

Similarly, from Eqs. (3.7a)–(3.8e) we find

$$\Psi_0(uH(u), vH(v)) = \Psi_0^{\mathrm{IV}}(uH(u), v) H(v) + \Psi_0^{\mathrm{Im}}(uH(u)) \delta(v),$$

$$\Psi_2(uH(u), vH(v)) = \Psi_2^{\mathrm{IV}}(u, v) H(u) H(v), \tag{3.24}$$

$$\Psi_4(uH(u), vH(v)) = \Psi_4^{\mathrm{IV}}(u, vH(v)) H(u) + \Psi_4^{\mathrm{Im}}(vH(v)) \delta(u),$$

and

$$\Phi_{00}(uH(u), vH(v)) = \Phi_{00}^{IV}(uH(u), v)H(v) + \Phi_{00}^{Im}(uH(u))\delta(v),$$

$$\Phi_{11}(uH(u), vH(v)) = \Phi_{11}^{IV}(u, v)H(u)H(v),$$

$$\Phi_{02}(uH(u), vH(v)) = \Phi_{02}^{IV}(u, v)H(u)H(v), \qquad (3.25)$$

$$\Phi_{22}(uH(u), vH(v)) = \Phi_{22}^{IV}(u, vH(v))H(u) + \Phi_{22}^{Im}(vH(v))\delta(u),$$

$$\Lambda(uH(u), vH(v)) = \Lambda^{IV}(u, v)H(u)H(v),$$

where $\Psi_i^{IV}(u, v)$ and $\Phi_{ij}^{IV}(u, v)$ are the corresponding Weyl and Ricci scalars in Region IV, obtained from Eqs. (3.7a)–(3.8e), and $\Psi_0^{Im} \ldots \Phi_{22}^{Im}$ are the impulsive part of the Riemann tensor with support on the hypersurfaces $u = 0$ and $v = 0$, respectively, and defined by

$$\Psi_0^{Im}(uH(u)) \equiv -\frac{1}{2}B^2(uH(u), 0)[V_{,v}(uH(u), 0) \cosh W(uH(u), 0)$$

$$- iW_{,v}(uH(u), 0)],$$

$$\Psi_4^{Im}(vH(v)) \equiv -\frac{1}{2}A^2(0, vH(v))[V_{,u}(0, vH(v)) \cosh W(0, vH(v)) \qquad (3.26)$$

$$+ iW_{,u}(0, vH(v))],$$

and

$$\Phi_{00}^{Im}(uH(u)) \equiv \frac{1}{2}B^2(uH(u), 0)U_{,v}(uH(u), 0),$$

$$\Phi_{22}^{Im}(vH(v)) \equiv \frac{1}{2}A^2(0, vH(v))U_{,u}(0, vH(v)). \qquad (3.27)$$

The terms Φ_{00}^{Im} and Φ_{22}^{Im} are usually interpreted as representing impulsive shells of null dust with support, respectively, on the hypersurfaces $v = 0$ and $u = 0$ (Dray and 't Hooft, 1986; Taub, 1988a; Tsoubelis, 1989b; Tsoubelis and Wang, 1990, 1991, 1992).

From the assumption that the metric coefficients are at least C^3 in the interaction region, we can see that $\Psi_i^{IV}(u, v)$ and $\Phi_{ij}^{IV}(u, v)$ are C^1 in that region, while Ψ_0^{Im}, Φ_{00}^{Im}, and Ψ_4^{Im}, Φ_{22}^{Im} are C^2, respectively, on the hypersurfaces $v = 0$, and $u = 0$. Then, from Eqs. (3.24) and (3.25) we can see that Ψ_2, Φ_{11}, Φ_{02} and Λ, in general, have step function discontinuities across either the hypersurface $u = 0$ or $v = 0$.

On the other hand, it can be shown that for any given C^1-test function $F(u, v)$ we have

$$\langle [1 - H(u)]H(u), F(u, v)\rangle = \int_\Omega F(u, v)[1 - H(u)]H(u)dV$$

$$= \int_{\Omega+u} F(u, v)[1 - H(u)]dV = 0,$$

$$\langle H^2(u), F(u.v)\rangle = \int_\Omega F(u, v)H(u)H(u)dV$$

$$= \int_{\Omega+u} F(u, v)H(u)dV = \int_{\Omega+u} F(u, v)dV,$$

$$\langle [1 - H(u)]^2, F(u, v)\rangle = \int_\Omega F(u, v)[1 - H(u)][1 - H(u)]dV$$

$$= \int_{\Omega-u} F(u, v)[1 - H(u)]dV$$

$$= \int_{\Omega-u} F(u, v)dV, \tag{3.28}$$

or equivalently

$$[1 - H(u)]H(u) = 0, \quad H^2(u) = H(u), \quad [1 - H(u)]^2 = [1 - H(u)], \tag{3.29}$$

in the sense of distributions. From Eqs. (3.19)–(3.20) and (3.29), we find

$$\Psi_4^{IV}(u > 0, vH(v)) = \Psi_4^{IV}(u > 0, v > 0)H(v) + \Psi_4^{IV}(u > 0, 0)[1 - H(v)],$$

$$\Psi_4^{Im}(vH(v)) = \Psi_4^{Im}(v > 0)H(v) + \Psi_4^{Im}(0)[1 - H(v)], \tag{3.30}$$

etc., where

$$\Psi_4^{IV}(u > 0, 0) = \lim_{v \to 0^+} \Psi_4^{IV}(u > 0, v),$$

$$\Psi_4^{Im}(0) = \lim_{v \to 0^+} \Psi_4^{Im}(v). \tag{3.31}$$

Since $\Psi_4^{IV}(u > 0, v > 0)$ and $\Psi_4^{Im}(v > 0)$ are at least C^1 and C^2, respectively, we can see that $\Psi_4^{IV}(u > 0, vH(v))$ and $\Psi_4^{Im}(vH(v))$ are C^0 across the hypersurface $v = 0$.

Similarly, it can be shown that $\Phi_{22}^{IV}(u > 0, vH(v))$ and $\Phi_{22}^{Im}(vH(v))$ are C^0, too, across the hypersurface $v = 0$, while $\Psi_0^{IV}(uH(u), v > 0)$,

$\Psi_0^{Im}(uH(u))$, $\Phi_{00}^{IV}(uH(u), v > 0)$ and $\Phi_{00}^{Im}(uH(u))$ are C^0 across the hypersurface $u = 0$. Thus, we have

$$\langle D[\Psi_4(uH(fu), vH(v))], F(u,v)\rangle$$

$$\equiv \langle l^\mu \partial_\mu [\Psi_4(uH(u), vH(v))], F(u,v)\rangle$$

$$= \langle B\partial_v [\Psi_4(uH(u), vH(v))], F(u,v)\rangle$$

$$= \langle B\partial_v \{\Psi_4^{IV}(u,v)H(u)H(v) + \Psi_4^{IV}(u,0)H(u)[1 - H(v)]$$

$$+ \Psi_4^{Im}(v)H(v)\delta(u) + \Psi_4^{Im}(0)[1 - H(v)]\delta(u)\}, F(u,v)\rangle$$

$$= \langle H(u)H(v)B\partial_v [\Psi_4^{IV}(u,v)] + H(v)\delta(u)B\partial_v [\Psi_4^{Im}(v)], F(u,v)\rangle$$

$$\langle D[\Psi_2(uH(u), vH(v))], F(u,v)\rangle$$

$$= \langle D[\Psi_2^{IV}(u,v)H(u)H(v)], F(u,v)\rangle$$

$$= \langle H(u)H(v)D[\Psi_2^{IV}(u,v)]$$

$$+ \Psi_2^{IV}(u,v)H(u)DH(v), F(u,v)\rangle, \tag{3.32}$$

or equivalently

$$D[\Psi_4(uH(u), vH(v))] = H(u)H(v)D[\Psi_4^{IV}(u,v)] + H(v)\delta(u)D[\Psi_4^{Im}(v)],$$

$$D[\Psi_2(uH(u), vH(v))] = H(u)H(v)D[\Psi_2^{IV}(u,v)] + BH(u)\delta(v)\Psi_2^{IV}(u,0). \tag{3.33}$$

Substituting Eq. (3.23) into Eqs. (3.9a)–(3.10b), and taking Eq. (3.33) into account, we finally obtain

$$\Delta\Psi_0^{IV} + D\Phi_{02}^{IV} = (4\gamma - \mu)\Psi_0^{IV} + 3\sigma\Psi_2^{IV} + (2\varepsilon - 2\bar\varepsilon + \rho)\Phi_{02}^{IV}$$

$$+ 2\sigma\Phi_{11}^{IV} - \bar\lambda\Phi_{00}^{IV},$$

$$3D\Psi_2^{IV} - 2D\Phi_{11}^{IV} + \Delta\Phi_{00}^{IV} = -3\lambda\Psi_0^{IV} + 9\rho\Psi_2^{IV} + (\mu + 2\gamma + 2\bar\gamma)\Phi_{00}^{IV}$$

$$- 2\rho\Phi_{11}^{IV} - 2\sigma\Phi_{20}^{IV} + \bar\sigma\Phi_{02}^{IV},$$

$$3\Delta\Psi_2^{IV} - 2\Delta\Phi_{11}^{IV} + D\Phi_{22}^{IV} = 3\sigma\Psi_4^{IV} - 9\mu\Psi_2^{IV} - (\rho + 2\varepsilon + 2\bar\varepsilon)\Phi_{22}^{IV}$$

$$+ 2\mu\Phi_{11}^{IV} + 2\lambda\Phi_{02}^{IV} - \bar\lambda\Phi_{20}^{IV},$$

$$\tag{3.34}$$

$$D\Psi_4^{IV} + \Delta\Phi_{20}^{IV} = (\rho - 4\varepsilon)\Psi_4^{IV} - 3\lambda\Psi_2^{IV} - (\mu + 2\gamma - 2\bar\gamma)\Phi_{20}^{IV},$$

$$- 2\lambda\Phi_{11}^{IV} + \bar\sigma\Phi_{22}^{IV}, \quad (u, v > 0),$$

and

$$D\Phi_{11}^{IV} + \Delta\Phi_{00}^{IV} + 3D\Lambda^{IV} = 2(\gamma + \overline{\gamma} - \mu)\Phi_{00}^{IV} + 4\rho\Phi_{11}^{IV}$$
$$+ \overline{\sigma}\Phi_{02}^{IV} + \sigma\Phi_{20}^{IV},$$

$$\Delta\Phi_{11}^{IV} + D\Phi_{22}^{IV} + 3\Delta\Lambda^{IV} = 2(\rho - \varepsilon - \overline{\varepsilon})\Phi_{22}^{IV} - 4\mu\Phi_{11}^{IV}$$
$$- \overline{\lambda}\Phi_{20}^{IV} - \lambda\Phi_{02}^{IV}, \quad (u, v > 0),$$

$$(3.35)$$

in the interacting region (Region IV),

$$\Delta\Psi_0^{Im} + B\Phi_{02}^{IV} = (4\gamma - \mu)\Psi_0^{Im} - \overline{\lambda}\Phi_{00}^{Im},$$

$$\Delta\Phi_{00}^{Im} + 3B\Psi_2^{IV} - 2B\Phi_{11}^{IV} = (\mu + 2\gamma + 2\gamma)\Phi_{00}^{Im} - 3\lambda\Psi_0^{Im}, \qquad (3.36)$$

$$\Delta\Phi_{00}^{Im} + B\Phi_{11}^{IV} + 3B\Lambda^{IV} = 2(\gamma + \overline{\gamma} - \mu)\Phi_{00}^{Im}, \quad (u > 0, v = 0),$$

along the null surface $v = 0$ but with $u > 0$, and

$$D\Psi_4^{Im} + A\Phi_{20}^{IV} = -(4\varepsilon - \rho)\Psi_4^{Im} + \overline{\sigma}\Phi_{22}^{Im},$$

$$D\Phi_{22}^{Im} + 3A\Psi_2^{IV} - 2A\Phi_{11}^{IV} = -(\rho + 2\varepsilon + 2\overline{\varepsilon})\Phi_{22}^{Im} + 3\sigma\Psi_4^{Im}, \qquad (3.37)$$

$$D\Phi_{22}^{Im} + A\Phi_{11}^{IV} + 3A\Lambda^{IV} = -2(\varepsilon + \overline{\varepsilon} - \rho)\Phi_{22}^{Im}, \quad (u = 0, v > 0),$$

along the null surface $u = 0$ with $v > 0$. Inside Regions I–III, the Bianchi identities are satisfied identically.

Equation (3.34) shows that the shock parts of the two plane gravitational waves represented, respectively, by Ψ_0^{IV} and Ψ_4^{IV}, interact with each other through the Coulomb field Ψ_2^{IV}. The components Φ_{02}^{IV} and Φ_{11}^{IV} of the matter field interact with both Ψ_0^{IV} and Ψ_4^{IV}. That is, Φ_{02}^{IV} and Φ_{11}^{IV} are gravitationally active to both Ψ_0^{IV} and Ψ_4^{IV} (Szekeres, 1965). Similarly, Φ_{00}^{IV} is gravitationally active only to Ψ_0^{IV}, and Φ_{22}^{IV} only to Ψ_4^{IV}. The component Λ^{IV} is gravitationally inert to both Ψ_0^{IV} and Ψ_4^{IV}.

On the other hand, from Eqs. (3.36) and (3.37) we can see that the component Φ_{02}^{IV} is gravitationally active to both Ψ_0^{Im} and Ψ_4^{Im}, Φ_{00}^{Im} only to Ψ_0^{Im}, and Φ_{22}^{Im} only to Ψ_4^{Im}, while all the other components of the matter field are gravitationally inert to both Ψ_0^{Im} and Ψ_4^{Im}.

From Eqs. (3.36) and (3.37) we can also see that the interaction between purely gravitational impulsive waves is carried out through the Coulomb field Ψ_2^{IV} (Khan and Penrose, 1971; Nutku and Halil, 1977; Wang, 1991f,

1992a). Thus, it is necessary for the spacetime to be curved in the interaction region (Region IV) for the collision of two purely gravitational impulsive waves with support only on the hypersurfaces $u = 0$ and $v = 0$, although in the pre-collision regions (Regions I–III), the spacetime is flat.

3.3. Different Coordinate Systems

In the studies of spacetimes for colliding plane gravitational waves, several different coordinate systems are usually used. A successful choice of coordinate systems is of great importance both for solving the NP equations and for studying the properties of the solutions.

One of the most suitable coordinate system for colliding plane waves is the (u, v, x^2, x^3) coordinates, since in this system the two plane gravitational waves move along the $u = $ constant and $v = $ constant hypersurfaces, respectively, and the spacetime is automatically divided into four regions [see Fig. 3.1]. However, experience tells us that this coordinate system is not always convenient, especially when we try to find exact solutions of the Einstein field equations. In this section, we shall introduce several coordinate systems (Ferrari, 1988; Griffiths, 1991; Wang, 1991f), each of which meets our special need. Since the metric coefficients depend only on the null coordinates u and v, the choice of the coordinates x^2 and x^3 is the same in all these regions. Thus, in the following we consider only the different choices of the coordinates x^0 and x^1.

3.3.1. The (t, z)-coordinate system

The (t, z)-coordinates are commonly used to study the singularities formed after the collision (Yurtsever, 1987, 1988a, 1989; Feinstein and Ibañez, 1989; Griffiths, 1991), and to obtain exact solutions (Ferrari and Ibañez, 1987a, 1987b, 1989a, 1989b; Ferrari, Ibañez and Bruni, 1987a, 1987b; Wang, 1991a; Tsoubelis and Wang, 1992) by means of the soliton technique (or inverse scattering method) of Belinsky and Zakharov (Belinsky and Zakharov, 1978, 1979; Belinsky and Verdaguer, 2001).

The coordinates t and z are given by

$$t \equiv 1 - u^{2n} - v^{2m}, \quad z \equiv u^{2n} - v^{2m}, \tag{3.38}$$

or inversely

$$u = \sqrt[2n]{\frac{1 - t + z}{2}}, \quad v = \sqrt[2m]{\frac{1 - t - z}{2}}, \tag{3.39}$$

where n and m are two constants. In addition, in this coordinate system a special gauge is usually chosen

$$e^{-U} = t. \tag{3.40}$$

In terms of t and z, the metric (2.22) takes the form

$$ds^2 = f(t, z)(dt^2 - dz^2) - t\{e^V \cosh W (dx^2)^2 - 2\sinh W dx^2 dx^3$$
$$+ e^{-V} \cosh W (dx^3)^2\}, \tag{3.41}$$

where now V, W are functions of t and z, and the function $f(t, z)$ is defined by

$$f \equiv \frac{e^{-M}}{8nmu^{2n-1}v^{2m-1}}. \tag{3.42}$$

3.3.2. The (η, μ)-coordinate system

The (η, μ)-coordinate system was first used by Chandrasekhar and Ferrari (1984) for the spacetimes for colliding plane gravitational waves and later used by several others, see, for example, Tsoubelis and Wang (1989, 1992), Griffiths (1991), Wang (1991a), and references therein. In this coordinate system, the Einstein field equations for the functions V and W take the form

$$(1 - |E|^2)\{[(1 - \eta^2)E_{,\eta}]_{,\eta} - [(1 - \mu^2)E_{,\mu}]_{,\mu}\}$$
$$= -2\bar{E}[(1 - \eta^2)(E_{,\eta})^2 - (1 - \mu^2)(E,\mu)^2], \tag{3.43}$$

which was originally found by Ernst (1968a, 1968b), where the Ernst potential E is defined via the relations

$$Z \equiv \chi + iq_2, \quad E \equiv \frac{Z - 1}{Z + 1}, \quad \chi \equiv \frac{e^V}{\cosh W}, \quad q_2 \equiv \frac{\sinh W e^V}{\cosh W}, \tag{3.44}$$

and η, μ are given by

$$\eta = u^n X + v^m Y, \quad \mu = u^n X - v^m Y,$$
$$X = (1 - v^{2m})^{1/2}, \quad Y = (1 - u^{2n})^{1/2}. \tag{3.45}$$

Taking the first and second derivatives of η and μ with respect to u and v, we obtain the following useful relations:

$$\eta_{,u} = \frac{nu^{n-1}}{Y}(XY - u^n v^m), \quad \eta_{,v} = \frac{mv^{m-1}}{X}(XY - u^n v^m),$$

$$\eta_{,uu} = \frac{n(n-1)u^{n-2}}{Y}(XY - u^n v^m) - \frac{n^2 v^m u^{2n-2}}{Y^3},$$

$$\eta_{,vv} = \frac{m(m-1)v^{m-2}}{X}(XY - u^n v^m) - \frac{m^2 u^n v^{2m-2}}{X^3},$$

$$\eta_{,uv} = -\frac{nmu^{n-1}v^{m\cdot 1}}{XY}\eta, \tag{3.46}$$

$$\mu_{,u} = \frac{nu^{n-1}}{Y}(XY + u^n v^m), \quad \mu_{,v} = -\frac{mv^{m-1}}{X}(XY + u^n v^m),$$

$$\mu_{,uu} = \frac{n(n-1)u^{n-2}}{Y}(XY + u^n v^m) + \frac{n^2 v^m u^{2n-2}}{Y^3},$$

$$\mu_{,vv} = -\frac{m(m-1)v^{m-2}}{X}(XY + u^n v^m) - \frac{m^2 u^n v^{2m-2}}{X^3},$$

$$\mu_{,uv} = \frac{nmu^{n-1}v^{m-1}}{XY}\mu.$$

In addition, we also have

$$1 - \eta^2 = (XY - u^n v^m)^2, \quad 1 - \mu^2 = (XY + u^n v^m)^2. \tag{3.47}$$

Combining Eqs. (3.46) and (3.47), we find

$$\frac{d\eta^2}{1 - \eta^2} - \frac{d\mu^2}{1 - \mu^2} = \frac{(\eta_{,u}du + \eta_{,v}dv)^2}{1 - \eta^2} - \frac{(\mu_{,u}du + \mu_{,v}dv)^2}{1 - \mu^2}$$

$$= \frac{4nmu^{n-1}v^{m-1}}{XY}dudv. \tag{3.48}$$

On the other hand, from Eqs. (3.38) and (3.45) we obtain

$$t = (1 - \eta^2)^{1/2}(1 - \mu^2)^{1/2}, \quad z = \eta\mu. \tag{3.49}$$

Thus, we have

$$t_{,\eta} = -\frac{t\eta}{1 - \eta^2}, \quad t_{,\mu} = -\frac{t\mu}{1 - \mu^2}, \quad z_{,\eta} = \mu, \quad z_{,\mu} = \eta, \tag{3.50}$$

and

$$dt^2 - dz^2 = (\eta^2 - \mu^2)\left(\frac{d\eta^2}{1 - \eta^2} - \frac{d\mu^2}{1 - \mu^2}\right). \tag{3.51}$$

Substituting Eqs. (3.44), (3.49) and (3.51) into Eq. (3.41), we obtain

$$ds^2 = (\eta^2 - \mu^2)f(\eta,\mu)\left(\frac{d\eta^2}{1-\eta^2} - \frac{d\mu^2}{1-\mu^2}\right)$$

$$- [(1-\eta^2)(1-\mu^2)]^{1/2}[\chi(dx^2)^2 + \chi^{-1}(dx^3 - q_2 dx^2)^2]. \quad (3.52)$$

3.3.3. The (ϕ, θ)-coordinate system

To calculate the Weyl and Ricci scalars, it was found convenient to work in the ϕ and θ coordinates (Chandrasekhar and Ferrari, 1984; Chandrasekhar and Xanthopoulos, 1985a, 1986a), which are defined by

$$\eta = \cos\phi, \quad \mu = \cos\theta. \quad (3.53)$$

In this coordinate system, the metric (3.52) takes the form

$$ds^2 = (\cos^2\phi - \cos^2\theta)f(\phi,\theta)(d\phi^2 - d\theta^2)$$

$$- \sin\phi\sin\theta\{\chi(dx^2)^2 - \chi^{-1}(dx^3 - q_2 dx^2)^2\}. \quad (3.54)$$

It can be shown that the following relations hold:

$$u^n = \cos\frac{\phi+\theta}{2}, \quad v^m = \sin\frac{\theta-\phi}{2},$$

$$t = \sin\phi\sin\theta, \quad z = \cos\phi\cos\theta. \quad (3.55)$$

3.4. Gravitational Analog of the Faraday Rotation

In 1846, Faraday (1846a, 1846b) discovered that a magnetic field could affect the propagation of light. Suppose that a uniform magnetic field is imposed in a medium, and let a linearly polarized light beam propagate along the line of the force, then it was found that this linearly polarized light beam emerges with its plane of polarization rotated through an angle proportional to the magnetic field and the thickness of the medium. In electrodynamics, this phenomenon is known as *the Faraday rotation* (or Faraday effect). Because of the similarity between the electromagnetic and gravitational fields, we expect that a similar phenomenon also occurs in a gravitational field. Stark, Connors and Piran (Stark and Connors, 1977; Connors and Stark, 1977; Connors, Piran and Stark, 1980) first investigated the problem by observing the polarization vector of an X-ray emitted from an accretion disk surrounding a Kerr black hole, and found that the polarization vector rotates in the gravitational field of a black hole. Later,

Ishihara, Takahashi and Tomimatsu (1988) studied the gravitational lensing effect caused by a Kerr black hole via the propagation of a polarization vector along a light ray. In particular, they found a rotation of the plane of polarization due to the presence of the black hole's spin. The rotation angle is proportional to the mass and the line-of-sight component of the angle momentum of the black hole.

Later, Piran, Safier and Stark (Piran and Safier, 1985; Piran, Safier, and Stark, 1985; Piran, Safier, and Katz, 1986) studied the cylindrically symmetric spacetimes (Bronnikov, Santos and Wang, 2019). Starting from a cylindrical line element, they first defined the "+" and "×" modes of polarizations of a cylindrical gravitational wave, and then by using both analytic approximations and numerical calculations they showed that a conversion between the "+" and "×" modes occurs. If an outgoing (or ingoing) cylindrical wave is linearly polarized, its polarization vector rotates as it propagates. This is another gravitational analog of the electromagnetic Faraday rotation, but with cylindrical symmetry. It should be noted that the decomposition of the "+" and "×" modes considered by these authors is coordinate-dependent.

In this section, by using the definition given in Section 2.3, we shall provide a coordinate-independent analysis for the changes of polarizations, but now with plane symmetry (Wang, 1991b, 1991f).

As mentioned in Section 3.2, each of the null vectors l^μ and n^μ defines a null geodesic congruence. However, from Eq. (3.1) it can be seen that the space-like vectors $E^\mu_{(2)}$ and $E^\mu_{(3)}$ are parallelly transported along the null geodesics defined by n^μ or the ones by l^μ only in Regions II and III, but not in the interaction region (Region IV). As a matter of fact, now we have

$$
\begin{aligned}
E^\mu_{(2);\nu} l^\nu &= \frac{1}{2} \sinh W V_{,\nu} l^\nu E^\mu_{(3)}, \\
E^\mu_{(3);\nu} l^\nu &= -\frac{1}{2} \sinh W V_{,\nu} l^\nu E^\mu_{(2)},
\end{aligned}
\tag{3.56}
$$

and

$$
\begin{aligned}
E^\mu_{(2);\nu} n^\nu &= \frac{1}{2} \sinh W V_{,\nu} n^\nu E^\mu_{(3)}, \\
E^\mu_{(3);\nu} n^\nu &= -\frac{1}{2} \sinh W V_{,\nu} n^\nu E^\mu_{(2)}.
\end{aligned}
\tag{3.57}
$$

Thus, the angles ϕ_0 and ϕ_4 defined, respectively, by Eqs. (2.46) and (2.38) have physical meaning only locally in the interaction region (Region IV). In order to compare the polarization of a plane gravitational wave at different times and places, we have to find a parallelly transported basis along the wave path and then define the polarization angle relative to this parallelly transported basis. In this way, we can see that the changes of polarization of the wave has an absolutely physical meaning and independent of the choice of the coordinates. For example, if the changes (relative to the parallelly transported basis) are zero, it means that the polarization vector $E'^{\mu}_{(2)}$ defined in Section 2.3 is parallel to the parallelly transported basis. Clearly, such defined parallelism is independent of the coordinates.

To find a parallelly transported basis along the null geodesics defined by l^{μ}, we make a rotation in the $(E^{\mu}_{(2)}, E^{\mu}_{(3)})$-plane as we did in Eq. (2.32) but with the angle ϕ_4 now denoted by $\phi_4^{(0)}$, and $E'^{\mu}_{(2)}$ and $E'^{\mu}_{(3)}$ by $\lambda^{\mu}_{(2)}$ and $\lambda^{\mu}_{(3)}$, respectively. Then, we find

$$\lambda^{\mu}_{(2);\nu}l^{\nu} = \frac{1}{2}(\sinh W V_{,\nu} - 2\varphi^{(0)}_{4,\nu})l^{\nu}\lambda^{\mu}_{(3)},$$

$$\lambda^{\mu}_{(3);\nu}l^{\nu} = -\frac{1}{2}(\sinh W V_{,\nu} - 2\varphi^{(0)}_{4,\nu})l^{\nu}\lambda^{\mu}_{(2)}. \tag{3.58}$$

Therefore, if the angle $\phi_4^{(0)}$ is chosen so that

$$\sinh W V_{,v} - 2\varphi^{(0)}_{4,v} = 0, \tag{3.59}$$

the space-like orthogonal vectors $\lambda^{\mu}_{(2)}$ and $\lambda^{\mu}_{(3)}$ are parallelly transported along the null geodesics (or the Ψ_4-wave path) defined by l^{μ}, and the difference

$$\theta_4 \equiv \varphi_4 - \varphi_4^{(0)}, \tag{3.60}$$

defines the angle between the polarization direction of the Ψ_4-wave and the $\lambda^{\mu}_{(2)}$ direction.

Similarly, if the basis $(E^{\mu}_{(2)}, E^{\mu}_{(2)})$ is rotated so that the rotated angle $\varphi_0^{(0)}$ satisfies

$$\sinh W V_{,u} - 2\varphi^{(0)}_{0,u} = 0, \tag{3.61}$$

the vectors $\lambda^{\mu}_{(2)}$ and $\lambda^{\mu}_{(3)}$ are parallelly transported along the Ψ_0-wave path, and the angle

$$\theta_0 \equiv \varphi_0 - \varphi_0^{(0)}, \tag{3.62}$$

determines the polarization direction of the Ψ_4-wave relative to the $(\lambda^{\mu}_{(2)}, \lambda^{\mu}_{(3)})$-basis. From Eqs. (3.59) and (3.61), we can see that the angle $\phi_4^{(0)}(\phi_0^{(0)})$ is constant in Region III (Region II). Without loss of generality, we can choose $\phi_4^{(0)}(\phi_0^{(0)})$ to be zero in Region III (Region II), such that the $(\lambda^{\mu}_{(2)}, \lambda^{\mu}_{(3)})$-basis coincides with the $(E^{\mu}_{(2)}, E^{\mu}_{(3)})$-basis in these regions.

On the other hand, Eq. (3.24) shows that the Ψ_0- and Ψ_4-waves, in general, consist of two parts. We call the δ-function part the impulsive part, and the H-function part the shock part, although the latter still includes three different cases: smooth wave, shock wave and the wave with unbounded wavefront (Tsoubelis and Wang, 1989). The treatment for these three cases, however, is the same for the present problem, so we shall not distinguish them in the following discussions. We shall use θ_0^{sh} and θ_0^{Im} to denote the polarization angles for the shock and impulsive parts of Ψ_0, and θ_4^{sh} and θ_4^{Im} for the shock and impulsive parts of Ψ_4, respectively. Thus, in Region II we have $\theta_0^{\text{sh}} = \theta_0^{\text{sh}}(v)$ and $\theta_0^{\text{Im}} = $ constant. That is, in Region II the shock part in general is non-collinearly polarized, while the impulsive part is always collinearly polarized. Along the wave path, the two angles θ_0^{sh} and θ_0^{Im} do not change in this region

$$\theta_{0,u}^{\text{sh}} = 0 = \theta_{0,u}^{\text{Im}}. \tag{3.63}$$

Of cause, in Regions I and III the angles θ_0^{sh} and θ_0^{Im} vanish, since Ψ_0 vanishes there.

Similarly, for the Ψ_4-wave, we have

$$\theta_{4,u}^{\text{sh}} = 0 = \theta_{4,u}^{\text{Im}}, \tag{3.64}$$

in Region III, and θ_4^{sh} and θ_4^{Im} vanish in Regions I and II, since Ψ_4 vanishes in these two regions.

In the interaction region (Region IV), the situation is different. From the Bianchi identities (3.34)–(3.37), we find that

$$\begin{aligned}
A\Psi_{0,u}^{\text{IV}} = \frac{1}{2}\{&A[4(\ln B)_{,u} + U_{,u} - 2i\sinh WV_{,u}]\Psi_0^{\text{IV}} \\
&-3B(\cosh WV_{,v} - iW_{,v})\Psi_{2,v}^{\text{IV}} - 2B\Phi_{02,v}^{\text{IV}} \\
&+B(U_{,v} - 2i\sinh WV_{,v})\Phi_{02}^{\text{IV}} \\
&-2B(\cosh WV_{,v} - iW_{,v})\Phi_{11}^{\text{IV}} \\
&-A(\cosh WV_{,u} - iW_{,u})\Phi_{00}^{\text{IV}}\},
\end{aligned}$$

$$\begin{aligned}
B\Psi_{4,v}^{\text{IV}} = \frac{1}{2}\{&B[4(\ln A)_{,v} + U_{,v} + 2i\sinh WV_{,v}]\Psi_4^{\text{IV}} \\
&-3A(\cosh WV_{,u} + iW_{,u})\Psi_2^{\text{IV}} - 2A\Phi_{20,u}^{\text{IV}} \\
&+A(U_{,u} + 2i\sinh WV_{,u})\Phi_{20}^{\text{IV}} \\
&-2A(\cosh WV_{,u} + iW_{,u})\Phi_{11}^{\text{IV}} \\
&-B(\cosh WV_{,v} + iW_{,v})\Phi_{22}^{\text{IV}}\}, \quad (u, v > 0), \tag{3.65}
\end{aligned}$$

and

$$A\Psi_{0,u}^{\text{Im}} = \frac{1}{2}\{A[4(\ln B)_{,u} + U_{,u} - 2i\sinh WV_{,u}]\Psi_0^{\text{Im}}$$

$$-2B\Phi_{02}^{\text{IV}} - A(\cosh WV_{,u} - iW_{,u})\Phi_{00}^{\text{Im}}\}, \quad (u > 0, v = 0),$$

$$B\Psi_{4,v}^{\text{Im}} = \frac{1}{2}\{B[4(\ln A)_{,v} + U_{,v} + 2i\sinh WV_{,v}]\Psi_4^{\text{Im}}$$

$$-2A\Phi_{02}^{\text{IV}} - B(\cosh WV_{,v} - iW_{,v})\Phi_{22}^{\text{Im}}\}, \quad (u = 0, v > 0).$$

$$(3.66)$$

It is now convenient to introduce the following "scale-invariant" quantities via the relations (Szekeres, 1972; and Griffiths, 1976b),

$$\Psi_0 = B^2\Psi_0^{(0)}, \quad \Psi_2 = AB\ \Psi_2^{(0)}, \quad \Psi_4 = A^2\Psi_4^{(0)},$$

$$\Phi_{00} = B^2\Phi_{00}^{(0)}, \quad \Phi_{11} = AB\Phi_{11}^{(0)}, \quad \Phi_{02} = AB\Phi_{02}^{(0)}, \quad (3.67)$$

$$\Phi_{22} = A^2\Phi_{22}^{(0)}, \quad \Lambda = AB\Lambda^{(0)}.$$

Since from now on only the "scale invariant" terms are used, we shall drop all of the super-indices "0" from these quantities without causing any confusion. Then, substituting Eq. (3.67) into Eqs. (3.65) and (3.66), respectively, we find

$$\Psi_{0,u}^{\text{IV}} = \frac{1}{2}\{[U_{,u} - 2i\sinh WV_{,u}]\Psi_0^{\text{IV}}$$

$$-3(\cosh WV_{,v} - iW_{,v})\Psi_2^{\text{IV}} - 2\Phi_{02,v}^{\text{IV}}$$

$$+(U_{,v} - 2M_{,v} - 2i\sinh WV_{,v})\Phi_{02}^{\text{IV}}$$

$$-2(\cosh WV_{,v} - iW_{,v})\Phi_{11}^{\text{IV}}$$

$$-(\cosh WV_{,u} - iW_{,u})\Phi_{00}^{\text{IV}}\},$$

$$\Psi_{4,v}^{\text{IV}} = \frac{1}{2}\{[U_{,v} + 2i\sinh WV_{,v}]\Psi_4^{\text{IV}}$$

$$-3(\cosh WV_{,u} + iW_{,u})\Psi_2^{\text{IV}} - 2\Phi_{20,u}^{\text{IV}}$$

$$+(U_{,u} - 2M_{,u} + 2i\sinh WV_{,u})\Phi_{20}^{\text{IV}}$$

$$-2(\cosh WV_{,u} + iW_{,u})\Phi_{11}^{\text{IV}}$$

$$-(\cosh WV_{,v} + iW_{,v})\Phi_{22}^{\text{IV}}\}, \quad (u, v > 0),$$

$$(3.68)$$

and

$$\Psi_{0,u}^{\mathrm{Im}} = \frac{1}{2}\{[U_{,u} - 2i\sinh W V_{,u}]\Psi_0^{\mathrm{Im}}$$
$$-2\Phi_{02}^{\mathrm{IV}} - (\cosh W V_{,u} - iW_{,u})\Phi_{00}^{\mathrm{Im}}\}, \quad (u > 0, v = 0),$$

$$\Psi_{4,v}^{\mathrm{Im}} = \frac{1}{2}\{[U_{,v} + 2i\sinh W V_{,v}]\Psi_4^{\mathrm{Im}}$$
$$-2\Phi_{02}^{\mathrm{IV}} - (\cosh W V_{,v} - iW_{,v})\Phi_{22}^{\mathrm{Im}}\}, \quad (u > 0, v = 0). \tag{3.69}$$

Hence, we find that

$$\theta_{0,u}^{\mathrm{sh}} = -\frac{1}{4(\Psi_0^{\mathrm{IV}}\overline{\Psi_0^{\mathrm{IV}}})}\{3[\cosh W V_{,v}\,\mathrm{Im}(\Psi_0^{\mathrm{IV}}\overline{\Psi_2^{\mathrm{IV}}}) + W_{,v}\,\mathrm{Re}(\Psi_0^{\mathrm{IV}}\overline{\Psi_2^{\mathrm{IV}}})]$$
$$+ 2\,\mathrm{Im}(\Psi_0^{\mathrm{IV}}\Phi_{20,v}^{\mathrm{IV}}) + (2M_{,v} - U_{,v})\mathrm{Im}(\Psi_0^{\mathrm{IV}}\Phi_{20}^{\mathrm{IV}})$$
$$+ 2\Phi_{11}^{\mathrm{IV}}[\cosh W V_{,v}\,\mathrm{Im}(\Psi_0^{\mathrm{IV}}) + W_{,v}\,\mathrm{Re}(\Psi_0^{\mathrm{IV}})]$$
$$+ \Phi_{00}^{\mathrm{IV}}[\cosh W V_{,u}\,\mathrm{Im}(\Psi_0^{\mathrm{IV}}) + W_{,u}\,\mathrm{Re}(\Psi_0^{\mathrm{IV}})]$$
$$- 2\sinh W V_{,v}\,\mathrm{Re}(\Psi_0^{\mathrm{IV}}\Phi_{20}^{\mathrm{IV}})\}, \tag{3.70}$$

$$\theta_{4,v}^{\mathrm{sh}} = \frac{1}{4(\Psi_4^{\mathrm{IV}}\overline{\Psi_4^{\mathrm{IV}}})}\{3[\cosh W V_{,u}\,\mathrm{Im}(\Psi_4^{\mathrm{IV}}\overline{\Psi_2^{\mathrm{IV}}}) - W_{,u}\,\mathrm{Re}(\Psi_4^{\mathrm{IV}}\overline{\Psi_2^{\mathrm{IV}}})]$$
$$+ 2\,\mathrm{Im}(\Psi_4^{\mathrm{IV}}\Phi_{02,u}^{\mathrm{IV}}) + (2M_{,u} - U_{,u})Im(\Psi_4^{\mathrm{IV}}\Phi_{02}^{\mathrm{IV}})$$
$$+ 2\Phi_{11}^{\mathrm{IV}}[\cosh W V_{,v}\,\mathrm{Im}(\Psi_4^{\mathrm{IV}}) - W_{,v}\,\mathrm{Re}(\Psi_4^{\mathrm{IV}})]$$
$$+ \Phi_{22}^{\mathrm{IV}}[\cosh W V_{,v}\,\mathrm{Im}(\Psi_0^{\mathrm{IV}}) - W_{,v}\,\mathrm{Re}(\Psi_4^{\mathrm{IV}})]$$
$$+ 2\sinh W V_{,u}\,\mathrm{Re}(\Psi_4^{\mathrm{IV}}\Phi_{02}^{\mathrm{IV}})\}, \quad (u, v > 0),$$

and

$$\theta_{0,u}^{\mathrm{Im}} = -\frac{1}{4(\Psi_0^{\mathrm{Im}}\overline{\Psi_0^{\mathrm{Im}}})}\{[\cosh W V_{,u}\,\mathrm{Im}(\Psi_0^{\mathrm{Im}}) + W_{,u}\,\mathrm{Re}(\Psi_0^{\mathrm{Im}})]\Phi_{00}^{\mathrm{Im}}$$
$$+ 2\,\mathrm{Im}(\Psi_0^{\mathrm{Im}}\Phi_{20}^{\mathrm{IV}})\}, \quad (u > 0, v = 0),$$

$$\theta_{4,v}^{\mathrm{Im}} = \frac{1}{4(\Psi_4^{\mathrm{Im}}\overline{\Psi_4^{\mathrm{Im}}})}\{[\cosh W V_{,v}\,\mathrm{Im}(\Psi_4^{\mathrm{Im}}) - W_{,v}\,\mathrm{Re}(\Psi_4^{\mathrm{Im}})]\Phi_{22}^{\mathrm{Im}}$$
$$+ 2\,\mathrm{Im}(\Psi_4^{\mathrm{Im}}\Phi_{02}^{\mathrm{IV}})\}, \quad (u > 0, v = 0). \tag{3.71}$$

Equations (3.70) and (3.71) show that, due to the interaction between the two incoming plane gravitational waves and the interaction with the matter fields, the polarization directions of the Ψ_0- and Ψ_4-waves got changed relative to the parallelly transported basis along their wave paths.

In other words, the polarization direction of a plane gravitational wave is no longer parallelly transported along the wave path because of the above two kinds of interaction (Wang, 1991b). The change of polarizations of a plane gravitational wave caused by the interaction with another oppositely moving plane gravitational wave is the exact analog of the well-known electromagnetic Faraday rotation, but having the oppositely moving plane gravitational wave as the medium and the magnetic field. We call the effect caused by the nonlinear interaction between the plane gravitational waves and matter fields the deflection effect, and the effect caused by the interaction between the two plane gravitational waves the gravitational Faraday rotation (or gravitational Faraday effect).

When $W = 0$, all the Weyl and Ricci scalars are real [see Eqs. (3.7a)–(3.8e), (3.26) and (3.27)]. Consequently, from Eqs. (3.70) and (3.71) we find

$$\theta_{0,u}^{\text{sh}} = \theta_{4,v}^{\text{sh}} = \theta_{0,u}^{\text{Im}} = \theta_{4,v}^{\text{sh}} = 0, \quad (W = 0). \tag{3.72}$$

That is, in the collinear case the polarizations of colliding plane gravitational waves do not change.

To further illustrate the properties of the polarization of colliding plane gravitational waves, in the rest of this section we restrict ourselves to several specific cases (with $W \neq 0$), which are interesting from the point of view of physics. Because of the symmetry shared by the two plane gravitational waves, it is sufficient to consider only one of them, say, the Ψ_0-wave. In addition, since we are now working in Region IV, we do not make any more specific statements about it in the following, and understand all the following results valid only in this region (plus its two boundaries $u = 0, v > 0$ and $u > 0, v = 0$).

3.4.1. Vacuum spacetimes

When the spacetime is vacuum, the corresponding collision is a purely gravitational one, and the Ricci scalars are zero,

$$\Phi_{ij} = 0, \quad \Lambda = 0. \tag{3.73}$$

Then, Eq. (3.70) reduces to

$$\theta_{0,u}^{\text{sh}} = -\frac{3}{4(\Psi_0^{\text{IV}}\overline{\Psi_0^{\text{IV}}})}\{\cosh W V_{,v} \; \text{Im}(\Psi_0^{\text{IV}}\overline{\Psi_2^{\text{IV}}})$$
$$+ W_{,v} \; \text{Re}(\Psi_0^{\text{IV}}\overline{\Psi_2^{\text{IV}}})\}. \tag{3.74}$$

Thus, if $\Psi_2 = 0$, then we have $\theta_{0,u}^{\text{sh}} = 0$. That is, if $\Psi_2 = 0$, then the polarization of a plane gravitational wave does not change. However, it was shown that the "Coulomb" part Ψ_2 appears necessarily in the interaction region (Region IV), due to the nonlinearity of the Einstein field equations (Szekeres, 1972). Thus, *the change of polarizations of two colliding purely gravitational waves is purely due the nonlinear interaction between the two plane gravitational waves.*

On the other hand, from Eq. (3.68) we find

$$\theta_{0,u}^{\text{Im}}|_{v=0} = 0. \tag{3.75}$$

That is, *the impulsive plane gravitational wave does not change its polarization after collision, when no matter is present in the spacetime.* From Eq. (3.36) we can see that in the present case the Ψ_0^{Im}-wave component does not interact with other components.

3.4.2. Spacetimes filled with null dust

When a spacetime is filled with a null dust, the energy–stress tensor can be written in the form (Chandrasekhar and Xanthopoulos, 1986b; and Tsoubelis and Wang, 1990, 1991)

$$T_{\mu\nu} = \varepsilon_1 l_\mu l_\nu + \varepsilon_2 n_\mu n_\nu, \tag{3.76}$$

which is the superposition of two pure radiation fields given by Eq. (1.94), where ε_1 and ε_2 are non-negative. Equation (3.76) represents a pair of oppositely moving null dust clouds with the energy density ε_1 and ε_2, respectively, each of which is separately conserved (Taub, 1988a; Tsoubelis and Wang, 1991). The corresponding non-vanishing Ricci scalars are given by

$$\Phi_{00} = \frac{\varepsilon_2}{2B^2}, \quad \Phi_{22} = \frac{\varepsilon_1}{2A^2}. \tag{3.77}$$

On the other hand, Eq. (3.25) shows that, like Ψ_0 and Ψ_4, the components Φ_{00} and Φ_{22}, in general, consist of two parts: the H-function part and the δ-function part. The latter is supported on the hypersurfaces $u = 0$ or $v = 0$, and usually interpreted as impulsive shells of null dust. When attention is restricted to the inside of the interaction region, only the H-function part remains, and from Eq. (3.70) we find

$$\theta_{0,u}^{\text{sh}} = -\frac{1}{4(\Psi_0^{\text{IV}}\overline{\Psi_0^{\text{IV}}})}\{3[\cosh W V_{,v}\,\text{Im}(\Psi_0^{\text{IV}}\overline{\Psi_2^{\text{IV}}})$$

$$+ W_{,v}\,\text{Re}(\Psi_0^{\text{IV}}\overline{\Psi_2^{\text{IV}}})] + \Phi_{00}^{\text{IV}}[\cosh W V_{,u}\,\text{Im}(\Psi_0^{\text{IV}})$$

$$+ W_{,u}\,\text{Re}(\Psi_0^{\text{IV}})]\}. \tag{3.78}$$

Unlike in the vacuum case, now $\theta_{0,u}^{\text{sh}}$ can be different from zero even when $\Psi_2^{\text{IV}} = 0$, because of the presence of the last term in the right-hand side of Eq. (3.78), which represents the interaction between Ψ_0^{IV} and Φ_{00}^{IV} [see Eq. (3.34)].

It was shown that, when null dust is present the collision of two plane gravitational waves does not require the Coulomb field Ψ_2 to appear necessarily in the interaction region (Tsoubelis and Wang, 1990). Thus, a plane gravitational shock wave can change its polarization due to the deflection effect. On the other hand, Eq. (3.71) now becomes

$$\theta_{0,u}^{\text{Im}} = -\frac{1}{4(\Psi_0^{\text{Im}}\overline{\Psi_0^{\text{Im}}})}\{[\cosh W V_{,u}\,\text{Im}(\Psi_0^{\text{Im}})$$

$$+ W_{,u}\,\text{Re}(\Psi_0^{\text{Im}})]\Phi_{00}^{\text{Im}}\}. \tag{3.79}$$

Obviously, when Φ_{00}^{Im} is different from zero, the polarization of the impulsive part of Ψ_0 changes after collision because of the interaction between Ψ_0^{Im} and Φ_{00}^{Im} [see Eq. (3.36)].

3.4.3. Spacetimes filled with a massless scalar field

For a massless scalar field, the energy–stress tensor is given by Eq. (1.92) with the potential ϕ satisfying Eq. (1.93). Since in the present case all of the metric coefficients are functions of u and v only, without loss of generality, we assume that $\phi = \phi(u, v)$. Then, Eq. (1.93) becomes (Tsoubelis and Wang, 1991),

$$2\phi_{,uv} - U_{,u}\phi_{,v} - U_{,v}\phi_{,u} = 0. \tag{3.80}$$

The non-vanishing Ricci scalars are given by

$$\Phi_{00} = \frac{1}{2}\phi_{,v}^2, \quad \Phi_{22} = \frac{1}{2}\phi_{,u}^2,$$

$$\Phi_{11} = \frac{1}{4}\phi_{,u}\phi_{,v}, \quad \Lambda = -\frac{1}{12}\phi_{,u}\phi_{,v}. \tag{3.81}$$

Substituting Eq. (3.81) into Eq. (3.70), we find

$$\theta_{0,u}^{\text{sh}} = -\frac{1}{8(\Psi_0^{\text{IV}}\overline{\Psi_0^{\text{IV}}})}\{6[\cosh W V_{,v}\,\text{Im}(\Psi_0^{\text{IV}}\overline{\Psi_2^{\text{IV}}})$$

$$+ W_{,v}\,\text{Re}(\Psi_0^{\text{IV}}\overline{\Psi_2^{\text{IV}}})]$$

$$+ \phi_{,v}[\cosh W(\phi_{,u}\,V_{,v} + \phi_{,v}V_{,u})\text{Im}(\Psi_0^{\text{IV}})$$

$$+ (\phi_{,u}\,W_{,v} + \phi_{,u}W_{,v})\,\text{Re}(\Psi_0^{\text{IV}})]\}, \tag{3.82}$$

which shows that a plane gravitational shock wave changes its polarization due to both the nonlinear interaction between the two plane gravitational waves and the interaction with the massless scalar field ϕ.

When the spacetime is filled only with a massless scalar field, we have

$$\Phi_{00}^{\text{Im}} = 0 = \Phi_{22}^{\text{Im}}, \tag{3.83}$$

or equivalently

$$U_{,u}|_{u=0} = 0 = U_{,v}|_{v=0}. \tag{3.84}$$

Combining Eqs. (3.84) and (3.71) we find

$$\theta_{0,u}^{\text{Im}} = 0, \tag{3.85}$$

which means that an impulsive plane gravitational wave does not change its polarization when it passes through a massless scalar field, since in this case there is no interaction between the gravitational impulsive wave and the massless scalar field [see Eqs. (3.36) and (3.37)].

Note that Eq. (3.84) is also the condition under which the hypersurfaces $u = 0$ and $v = 0$ are free of matter (Tsoubelis and Wang, 1989, 1990).

3.4.4. Spacetimes filled with an electromagnetic field

When an electromagnetic field is null, its energy–stress tensor takes the form of a pure radiation field, which has been already discussed in Subsection 3.4.2. Thus, in this subsection we consider only the non-null case (Tsoubelis and Wang, 1991). Since the component ϕ_1 defined by Eq. (1.100) is zero in Regions II and III, from Eqs. (1.99a) and (1.99d) it can be shown that it must be also zero in Region IV. If we introduce the "scale invariant" quantities

$$\Phi_0 = B\Phi_0^{(0)}, \quad \Phi_2 = A\Phi_2^{(0)}, \tag{3.86}$$

and drop the superscript "0", then the non-vanishing Ricci scalars are

$$\Phi_{00}^{IV} = \Phi_0\overline{\Phi_0}, \quad \Phi_{02}^{IV} = \Phi_0\overline{\Phi_2} = \overline{\Phi_{20}^{IV}}, \quad \Phi_{22}^{IV} = \Phi_2\overline{\Phi_2}, \tag{3.87}$$

and the Maxwell field equations (1.99a)–(1.99d) reduce to

$$2\Phi_{0,u} = (U_{,u} - i\sinh W V_{,u})\Phi_0 - (\cosh W V_{,v} - iW_{,v})\Phi_2,$$
$$2\Phi_{2,u} = (U_{,v} + i\sinh W V_{,v})\Phi_2 - (\cosh W V_{,u} + iW_{,u})\Phi_0. \tag{3.88}$$

Note that in the present case all of the Ricci scalars have only the shock part, otherwise, the Maxwell potentials Φ_i (or equivalently, the electromagnetic field tensor $F_{\mu\nu}$) will contain the square roots of δ-function, which is not acceptable physically. Then, from Eqs. (3.70), (3.87) and (3.88) we find

$$\theta_{0,u}^{sh} = -\frac{1}{4(\Psi_0^{IV}\overline{\Psi_0^{IV}})}\{3[\cosh W V_{,v}\,\mathrm{Im}(\Psi_0^{IV}\overline{\Psi_2^{IV}})$$
$$+ W_{,v}\,\mathrm{Re}(\Psi_0^{IV}\overline{\Psi_2^{IV}})] + 2\,\mathrm{Im}(\Psi_0^{IV}\overline{\Phi}_{0,v}\Phi_2)$$
$$- \sinh W V_{,v}\,\mathrm{Re}(\Psi_0^{IV}\overline{\Phi}_0\Phi_2) + 2M_{,v}\,\mathrm{Im}(\Psi_0^{IV}\overline{\Phi}_0\Phi_2)\}. \tag{3.89}$$

Thus, similar to the last two cases, a plane gravitational shock wave can change its polarization when it interacts with an electromagnetic field.

On the other hand, for the impulsive part of Ψ_0, Eq. (3.71) becomes

$$\theta_{0,u}^{Im} = -\frac{\mathrm{Im}(\Psi_0^{Im}\overline{\Phi}_0\Phi_2)}{2(\Psi_0^{Im}\overline{\Psi_0^{Im}})}. \tag{3.90}$$

It follows that a gravitational impulsive wave can also change its polarization due to the interaction between the impulsive wave and the electromagnetic field component Φ_{02}^{IV}.

3.4.5. Spacetimes filled with a neutrino field

When a neutrino field is present, in general it requires the spin coefficients to be different from zero (Griffiths, 1976a, 1976b). Consequently, the metric for a neutrino field generally is not reducible to the simple form of Eq. (2.22) (Szekeres, 1972; Griffiths, 1976b; Tsoubelis and Wang, 1991). However, when one of the spinor components Ψ and Φ are zero [see Eq. (1.102)], the metric does take the form of Eq. (2.22) (Griffiths, 1976b; Tsoubelis and Wang, 1991). Then, Eq. (1.105) shows that the only non-vanishing Ricci scalar is either Φ_{00} (when $\Phi = 0$) or Φ_{22} (when $\Psi = 0$). But this corresponds to the case for a pure radiation field, which we have already discussed in Subsection 3.4.2.

3.5. Singularities in Spacetimes of Colliding Plane Gravitational Waves

The study of colliding plane gravitational waves was initially motivated by Penrose's conjecture that the focusing effect of single plane waves should cause the colliding waves to interact strongly and eventually produce space-time singularities (Penrose, 1965). Later on this conjecture was verified by several exact solutions (Szekeres, 1970, 1972; Khan and Penrose 1971; Nutku and Halil, 1977) and various theorems (Tippler, 1980; Centrella and Matzner, 1982; Matzner and Tippler, 1984). Thus, until 1986 it was believed that the collision of plane gravitational waves inevitably produced singularities in the future of the interaction. However, it were Chandrasekhar and Xanthopoulos who first pointed out that such singularities were not imperative to occur, and, instead, that Killing–Cauchy (or simply Cauchy) horizons may take place (Chandrasekhar and Xanthopoulos, 1986a). In parallel to the Chandrasekhar–Xanthopoulos work, Ferrari and Ibañez (1987a, 1988) found that this was also the case when the two plane gravitational waves are collinearly polarized. Following Chandrasekhar and Xanthopoulos, and Ferrari and Ibañez, the studies of singularities in the colliding plane gravitational wave spacetimes have attracted a great deal of attention. Yurtsever (1988a) first found that there exists an abundance of exact colliding collinearly polarized plane gravitational wave solutions, which create Cauchy horizons instead of spacetime curvature singularities. It was also shown that those Cauchy horizons are not stable in the full nonlinear theory against small but "generic" perturbations of the initial data,[2] and that "generic" initial data always produce space-like curvature singularities without Cauchy horizons. Tsoubelis and Wang (1989) and Feinstein and Ibañez (1989) obtained the same results by considering various solutions. Later, Yurtsever (1989) generalized his previous studies to cover the non-collinear case, and found that similar results also hold. Specifically, it was shown that at a fixed value of the space-like coordinate z [see Eq. (3.41)] the metric for

---

[2]A rigorous proof of the instability for the case in which the initial waves have constant aligned polarizations had to wait until 2005, when Griffiths first showed that such horizons are indeed unstable with respect to bounded perturbations of the initial waves (Griffiths, 2005). However, such a proof is still absent for the non-collinear case.

vacuum solutions takes the form of a "generalized inhomogeneous" Kasner vacuum solution,

$$ds^2 \approx \varepsilon_0(z)d\tau^2 - \varepsilon_1(z)\tau^{2p_1(z)}dz^2$$
$$- \varepsilon_2(z)\tau^{2p_2(z)}(dX_{(z)}^2)^2 - \varepsilon_3(z)\tau^{2p_3(z)}(dX_{(z)}^3)^2, \qquad (3.91)$$

as $t \to 0$ in both of the collinear and non-collinear cases, where the exponents $P_a(z)$ satisfy the Kasner relations

$$p_1(z) + p_2(z) + p_3(z) = p_1(z)^2 + p_2(z)^2 + p_3(z)^2 = 1, \qquad (3.92)$$

where τ is monotonically related to the time-like coordinate t, $\varepsilon_\mu(z)$ are regular functions at the fixed point z, and $X_{(z)}^2$ and $X_{(z)}^3$ are in general asymptotically z-dependent linear combinations of the coordinates x^2 and x^3. The exponent $p_1(z)$ is always referred to the z-axis, while $p_2(z)$ and $p_3(z)$ are referred to the asymptotic Kasner axes $X_{(z)}^2$ and $X_{(z)}^3$, respectively. Moreover, it was further shown that the asymptotic Kasner nature of the metric implies the corresponding asymptotic Kasner behavior of the spacetime curvature, which in turn implies that the solutions that are free of curvature singularities on the "focusing hypersurface" are the ones that their asymptotic Kasner form is that of a degenerate Kasner solution. It can be shown that the latter case always corresponds to

$$p_1(z) = 0. \qquad (3.93)$$

On the other hand, Tsoubelis and Wang studied the effect of polarizations of colliding plane gravitational waves on the formation and nature of singularities (Tsoubelis and Wang, 1992). Specifically, it was found that some astigmatic singularities are turned into anastigmatic singularities, or vice versa, due to the interaction between different polarization modes of colliding plane gravitational waves. Moreover, it was also found that all of the solutions that are free of singularities in the collinear case are so in the non-collinear case, but inversely not. That is, due to the interaction between different polarization modes some singularities are turned into Cauchy horizons.

In addition, Chandrasekhar and Xanthopoulos (1987a) studied the effects of sources on the Cauchy horizons, and found that the coupling of gravitational waves with an electromagnetic field does not affect in any essential way the development of the Cauchy horizons, if the polarizations of the colliding gravitational waves are not parallel. However, if the polarizations are parallel, the space-like singularity that occurs in the vacuum is transformed into a horizon followed by a three-dimensional time-like singularity by the merest presence of the electromagnetic field. In addition, the couplings of the gravitational waves with a "stiff" fluid and null dust affect the development of Cauchy horizons and singularities in radically different ways: the "stiff" fluid affects the development decisively in all cases but qualitatively in the same way, while null dust prevents the development of Cauchy horizons and allows only the development of space-like singularities. Tsoubelis and Wang (1991), on the other hand, showed that when the null dust is confined on null hypersurfaces, the conclusion for the null dust is quite different from the one given above, and found that the collision of plane gravitational waves does not necessarily develop a space-like curvature singularity in the interaction region, when coupled with null dust shells. Except the ones developed after the collision, singularities also form in the pre-collision regions. The latter are usually attributed to the "back-reaction", and have not been well understood so far. Later, Konkowski and Helliwell (1989, 1991) studied three typical cases and found that in the colliding impulsive wave spacetimes and colliding sandwich wave spacetimes the singularities in the pre-collision regions are quasi-irregular, and that in the colliding thick gravitational wave spacetimes they are non-scalar curvature singularities. But, they argued that all these singularities are not stable, and that the presence of matter or matter fields will convert these singularities into scalar curvature ones.

3.6. Methods for Generating New Solutions

The Einstein field equations with a tow-dimensional group G_2 of isometries acting orthogonally transitively on non-null orbits are nonlinear partial differential equations in two independent variables (Kramer *et al.*, 1980; Stephani *et al.*, 2009). When the orbits are time-like, the corresponding Einstein field equations are elliptical and the spacetimes are stationary axisymmetric, whereas when the orbits are space-like, the corresponding Einstein field equations are hyperbolic and the spacetimes are either cylindrically or plane symmetric. The plane symmetric case includes cosmological models as well as colliding plane gravitational waves.

Since the pioneering work of Geroch (1971, 1972), a fair amount of labor has been devoted to the above cases, and several methods have been developed and applied to the construction and analysis of properties of solutions for these corresponding partial differential equations (MacCallum, 1984; Stephani *et al.*, 2009). Among these methods are the Harrison Bäcklund transformations (Harrison, 1978, 1980), the Hoenselaers–Kinnersley–Xanthopoulos transformations (Hoenselaers Kinnersley and Xanthopoulos, 1979a, 1979b; Xanthopoulos, 1979, 1981), the Neugebauer–Kramer involution (Neugebauer, 1979, 1980a, 1980b; Kramer and Neugebauer, 1980, 1984), the Hauser–Ernst formulation of the Riemann–Hilbert problem (Hauser and Ernst, 1979a, 1979b, 1980a, 1980b, 1981; Palenta and Meinel, 2017), the inverse scattering method (or soliton technique) of Belinsky and Zakharov (Belinsky and Zakharov, 1978, 1979; Belinsky and Verdaguer, 2001), the Chandrasekhar–Ferrari method (Chandrasekhar and Ferrari, 1984), and more recently the initial value problem of Alekseev and Griffiths for colliding gravitational and electromagnetic waves (Alekseev and Griffiths, 2001, 2004). See also Griffiths and Santano-Roco (2002) for the initial value problem of colliding collinearly polarized plane gravitational waves.

The usefulness of the above-mentioned methods is mostly restricted to the vacuum case or to the case in which the Ricci tensor vanishes on the subspace spanned by the Killing vectors (Belinsky, 1979, 1980; Kitchingham, 1986).

Progress in the direction of relaxing the above-mentioned restrictions started with the work of Belinsky and Ruffini (1980). They first extended the inverse scattering method (ISM) of Belinsky and Zakharov (BZ) to the five-dimensional vacuum stationary axisymmetric case and showed that, through the Jordan–Kaluza–Klein dimensional reduction procedure, their results are relevant to the construction of exact solutions corresponding to a scalar–vector–tensor theory. Later, following Belinsky and Ruffini's approach, Ibañez and Verdaguer (1986) constructed some cosmological solutions representing solitonic perturbations of the Friedmann–Lemaître–Robertson–Walker (FLRW) four-dimensional background with an effective ultra-relativistic equation of state for the matter content. Diaz, Gleiser and Pullin (1987, 1988) further extended ISM of BZ to a class of perfect fluid solutions. As an example, they obtained finite perturbations of the solitonic type for the FLRW flat spacetime with a perfect fluid satisfying the "gamma" law of Eq. (1.108) for the equation of state. Yet, motivated by the five-dimensional representation of the Brans–Dicke–Jordan theory, Bruckman (1986, 1987) extended ISM (BZ) to D-dimensional stationary axisymmetric spacetimes.

In this section, we concern ourselves only with ISM (BZ) and Chandrasekhar–Ferrari method in the vacuum case for both their utility and simplicity. For the extended ISM (BZ), we refer the readers to the above-mentioned papers, the review presented by Verdaguer (1984) and the monograph of Belinsky and Verdaguer (2001). For other methods, the original works cited above will be the excellent references to look for.

3.6.1. The Chandrasekhar–Ferrari method

The Chandrasekhar–Ferrari method (Chandrasekhar and Ferrari, 1984) is based on the reciprocal relations that exist between stationary axisymmetric spacetimes and plane symmetric spacetimes. In particular, Chandrasekhar and Ferrari showed that in vacuum the metrics for plane symmetric spacetimes satisfy the same Ernst equation (3.43) as do the metrics for stationary axisymmetric spacetimes. Therefore, a correspondence exists between solutions of these two classes of spacetimes in the sense that they can be obtained from the same Ernst potential. By using this correspondence Chandrasekhar and Ferrari showed that the Schwarzschild and Kerr solutions correspond, respectively, to the Khan–Penrose (1971) and Nutku–Halil (1977) solutions. Following the same line, Chandrasekhar and Xanthopoulos have obtained many new solutions for either colliding purely gravitational plane waves (Chandrasekhar and Xanthopoulos, 1986a), or for colliding plane gravitational waves coupled with electromagnetic field (Chandrasekhar and Xanthopoulos, 1985a, 1987b), or for colliding plane gravitational waves coupled with null dust (Chandrasekhar and Xanthopoulos, 1986b), or for colliding plane gravitational wave coupled with a "stiff" fluid (Chandrasekhar and Xanthopoulos, 1985b, 1985c).

3.6.2. Inverse scattering method of Belinsky and Zakharov

When ISM (BZ) is used to obtain new (exact) solutions of the Einstein vacuum equations with plane symmetry, the metric is written in the form of Eq. (3.41). The corresponding vacuum equations are given by (Carr and Verdaguer, 1984; Verdaguer, 1984),

$$A_{,t} - B_{,z} = 0, \tag{3.94}$$

and

$$(\ln f)_{,t} = -\frac{1}{t} + \frac{1}{4t} \operatorname{Tr}(A^2 + B^2),$$

$$(\ln f)_{,z} = \frac{1}{2t} \operatorname{Tr}(AB), \tag{3.95}$$

where the matrices $A, B,$ and g are defined, respectively, by

$$g = t \begin{pmatrix} e^V \cosh W & -\sinh W \\ -\sinh W & e^{-V} \cosh W \end{pmatrix}, \tag{3.96}$$

$$A \equiv tg_{,t}g^{-1}, \quad B = tg_{,z}g^{-1}, \tag{3.97}$$

where $\mathrm{Tr}(Y)$ denotes the trace of the indicated matrix Y, and g^{-1} is the inverse of g.

It can be shown that the integrability conditions for Eq. (3.95) are automatically satisfied if the matrix g is subject to Eq. (3.94). Thus, once a solution of g is found, the function f can be obtained by integrating out Eq. (3.95) directly.

The main idea of ISM (BZ) is to look for a linear eigenvalue problem that has the nonlinear equations (3.94) as the integrability conditions. Solving the linear problem will produce, by an appropriate procedure, solutions to the original nonlinear equations.

The linear problem associating to Eq. (3.94) is

$$\left(\partial_t - \frac{2\lambda t}{\lambda^2 - t^2}\partial_\lambda\right)\psi = -\frac{tA + \lambda B}{\lambda^2 - t^2}\psi,$$
$$\left(\partial_z - \frac{2\lambda^2}{\lambda^2 - t^2}\partial_\lambda\right)\psi = -\frac{tB + \lambda A}{\lambda^2 - t^2}\psi, \tag{3.98}$$

where λ is a complex "spectral" parameter, and $\psi = \psi(\lambda, t, z)$ is a 2×2 matrix which satisfies the "initial condition"

$$\psi(\lambda = 0, t, z) = g(t, z). \tag{3.99}$$

Equations (3.97) and (3.98) consist of a genuine linear differential equation system for the unknown "wave-function" ψ with Eq. (3.99) as the initial condition.

Suppose that $\psi^{(0)}$ is a particular solution of Eqs. (3.97)–(3.99) for a given $g^{(0)}$ (the seed solution). Then, as Belinsky and Zakharov showed, the ansatz

$$\psi = X\psi^{(0)}, \tag{3.100}$$

gives a new solution of ψ, subject to the assumption that the "scattering" matrix $X(\lambda, t, z)$ has only single poles. By using the initial condition (3.99), we in turn obtain a new solution of g.

The key point in using ISM (BZ) is to integrate the differential equation system (3.98) to get an explicit solution $\psi^{(0)}$ for the given seed $g^{(0)}$. Then, the new solutions will be completely determined by the number and the nature (real or complex) of the poles. Each pole is characterized by the so-called pole trajectory given by

$$\mu_k = w_k - z \pm \sqrt{(w_k - z)^2 - t^2}, \quad (k = 1, 2, 3, \ldots, N), \tag{3.101}$$

where N is the number of poles, w_k are arbitrary (real or complex) parameters, and μ_k's are the solutions of the equations

$$\mu_{k,t} = \frac{2t\mu_k}{t^2 - \mu_k^2}, \quad \mu_{k,z} = \frac{2\mu_k^2}{t^2 - \mu_k^2}. \tag{3.102}$$

Then, the new solutions are given by

$$f = c_{ph}^{(0)} f^{(0)} t^{-N^2/2} \left(\prod_{k=1}^{N} |\mu_k|^{N+1} \right) \left[\prod_{l,k=1;k>l}^{N} (\mu_k - \mu_l)^{-2} \right] \det(\Gamma_{kl}),$$

$$g_{ab} = \left[\prod_{k=1}^{N} \left| \frac{\mu_k}{t} \right| \right] \left[g_{ab}^{(0)} - \sum_{j,k=1}^{N} \Gamma_{jk}^{-1} N_a^{(j)} N_b^{(k)} \right], \quad (a, b = 2, 3), \tag{3.103}$$

where

$$\Gamma_{jk} \equiv \sum_{a,b=2}^{3} \frac{n_a^{(j)} g_{ab}^{(0)} n_b^{(k)}}{\mu_j \mu_k - t^2},$$

$$N_a^{(j)} \equiv \sum_{b=2}^{3} \frac{n_b^{(k)} g_{ab}^{(0)}}{\mu_k} \quad \text{(without summation for the index } k\text{)}, \tag{3.104}$$

$$n_a^{(j)} \equiv \sum_{b=2}^{3} m_b^{(k)} M_{ab}^{(k)} \quad \text{(without summation for the index } k\text{)},$$

$$M^{(k)} \equiv [\psi^{(0)}(\mu_k, t, z)]^{-1},$$

where $C_{ph}^{(0)}$ is an arbitrary real constant, and $m_b^{(k)}$'s are arbitrary real or complex parameters. If we start with real poles, then the parameters $m_b^{(k)}$ also have to be real. If we start with a complex pole μ_k, its complex conjugate is also a pole. Thus, complex poles always go in pairs, and in this case we can set $\mu_{k+N/2} = \bar{\mu}_k$. The complex parameters $m_b^{(k)}$ will then satisfy the relations

$$m_a^{(k+N/2)} = \bar{m}_a^{(k)}. \tag{3.105}$$

It must be noted that the term in the square bracket in the first equation of Eq. (3.103) should be replaced by unit when $N = 1$.

From the above general description of ISM (BZ), we can see that a complete solution is obtained only after the system (3.98) is explicitly solved for a given seed $g^{(0)}$. However, as noticed by various authors, this in general is not trivial. The solutions for $\Psi^{(0)}$ were found for several simple diagonal seeds (Jantzen, 1980) and for a few non-diagonal seeds (Belinsky and Francaviglia, 1982; Letelier, 1986, 1989; Kitchingham, 1986; Cespedes and Verdaguer, 1987). Therefore, some further simplifications are usually adopted. A common assumption is that the matrices $\Psi^{(0)}$ and $g^{(0)}$ are diagonal. With the latter assumption, Economou and Tsoubelis (1989; see also Letelier, 1986) found that the corresponding expressions of Eqs. (3.103) and (3.104) can be written in a very simple form in terms of the determinants of four $N \times N$ matrices. The Economou–Tsoubelis results come from the following observations. Integrating the trace of Eq. (3.98) along a pole trajectory μ_k, we have

$$\det(\psi^{(0)}) = 2w_k\mu_k \quad \text{(without summation for the index } k\text{)}, \qquad (3.106)$$

which allows us to write the diagonal matrix $\psi^{(0)}$ in the form

$$\psi^{(0)}(\lambda, t, z) = \text{diag}\{\psi_k, 2w_k\mu_k\psi_k^{-1}\}$$

$$\text{(without summation for the index } k\text{)}, \qquad (3.107)$$

where ψ_k satisfies the equations,

$$(\ln \psi_k)_{,\zeta} = \frac{t}{t - \mu_k}(\ln g_{22}^{(0)})_{,\zeta}, \qquad (3.108)$$

$$(\ln \psi_k)_{,\xi} = \frac{t}{t + \mu_k}(\ln g_{22}^{(0)})_{,\xi}, \qquad (3.109)$$

with

$$\zeta \equiv t + z, \quad \xi = t - z. \qquad (3.110)$$

Introducing the following quantities:

$$\sigma_k \equiv \frac{\mu_k}{t}, \quad s_k \equiv \frac{Q_k g_{22}^{(0)}}{\psi_k^2}\sigma_k \quad \text{(without summation for the index } k\text{)},$$

$$\Delta_{jk} \equiv \frac{s_j s_{k+1}}{\sigma_j \sigma_k - 1} \quad \text{(without summation for the indexes } j \text{ and } k\text{)},$$

$$(3.111)$$

where Q_k are arbitrary constants, Economou and Tsoubelis found that the new solutions can be written in the form

$$g_{22} = \left[\prod_{k=1}^{N} |\sigma_k| \right] L_{(-1)} L_{(0)}^{-1} g_{22}^{(0)},$$

$$g_{23} = t \left[\prod_{k=1}^{N} |\sigma_k| \right] L L^{-1}(0),$$

$$g_{33} = \left[\prod_{k=1}^{N} |\sigma_k|^{-1} |L_{(+1)} L^{-1}(0) g_{22}^{(0)}, \right. \tag{3.112}$$

$$f = C_{ph}^{(0)} f^{(0)} L_{(0)} t^{-N^2/2} \left(\prod_{k=1}^{N} |\sigma_k|^N \right) \left[\prod_{l,k=1;\ k>l}^{N} (\sigma_k - \sigma_1)^{-2} \right]$$

$$\times \left(\prod_{k=1}^{N} |s_k|^{-1} \right),$$

where

$$L_{(d)} \equiv \det \Delta_{(d)}, \quad \Delta_{(d)jk} \equiv \frac{(\sigma_j \sigma_k)^d s_j s_k + 1}{\sigma_j \sigma_k - 1}, \quad (d = 0, \pm 1),$$

$$L \equiv \det \Delta_{(0)} - \det \left[\Delta_{(0)jk} + \frac{s_j}{\sigma_j \sigma_k} \right]. \tag{3.113}$$

Later, Tsoubelis and Wang found (Wang, 1991a; Tsoubelis and Wang, 1992) that the integration of Eq. (3.108) will become simpler, if we work in the coordinates η and μ introduced in Section 3.3. Actually, introducing the quantity $\Sigma^{(0)}$ via the relation

$$\psi^{(0)}(\lambda, t, z) = (2w\lambda)^{1/2} \operatorname{diag}\{\Sigma^{(0)}, \Sigma^{(0)-1}\}, \tag{3.114}$$

it was found that the system (3.98) (not restricted only along the pole trajectories) reduces to

$$[(1 - \eta^2)\partial_\eta + \lambda\partial_\mu - 2\lambda\eta\partial_\lambda] \ln \Sigma^{(0)} = (1 - \eta^2) V^{(0)},_\eta,$$
$$[(1 - \mu^2)\partial_\mu + \lambda\partial_\eta - 2\lambda\mu\partial_\lambda] \ln \Sigma^{(0)} = (1 - \mu^2) V^{(0)},_\mu, \tag{3.115}$$

and the "initial condition" of Eq. (3.115) becomes

$$\Sigma^{(0)}(\lambda = 0, t, z) = e^{V^{(0)}}. \tag{3.116}$$

The integrability conditions for Eq. (3.115) is the Einstein vacuum field equation for the seed function $V^{(0)}$,

$$[(1 - \eta^2) V^{(0)},_\eta],_\eta - [(1 - \mu^2) V^{(0)},_\mu],_\mu = 0, \quad (W^{(0)} = 0). \tag{3.117}$$

Chapter 4

Collision of Pure Gravitational Plane Waves

In this chapter, we shall present three classes of exact solutions of the Einstein vacuum field equations, which represent collisions of two purely gravitational plane waves. Specifically, in Section 4.1 we present a three-parameter class of diagonal solutions, which represents a variety of models of collision of two collinearly polarized gravitational plane waves. The main properties of these solutions are investigated. In Section 4.2, one of the above two colliding constantly polarized gravitational plane waves is generalized to a non-collinearly polarized one by adding one soliton into the above diagonal solutions. The resulting four-parameter class of one-soliton solutions in general represents the collision of a collinearly polarized and a non-collinearly polarized gravitational plane waves. The effects of the polarization of colliding plane gravitational waves on the formation and nature of singularities in the interacting region are studied. Finally, in Section 4.3, a five-parameter class of two-soliton solutions is presented, which include almost all the known (both diagonal and non-diagonal) solutions of colliding gravitational plane waves. The reciprocal relations between these two-soliton solutions and the ones for stationary axisymmetric case are also investigated. The effects of polarizations of these colliding gravitational waves on the formation and nature of spacetime singularities due to the mutual focusing of the two colliding purely gravitational waves are discussed.

4.1. Collisions of Collinearly Polarized Gravitational Plane Waves

The Einstein vacuum equations for colliding plane gravitational waves can be obtained from Eqs. (3.8a)–(3.8e) by simply setting all the Ricci scalars zero

$$\Phi_{ij} = 0, \qquad \Lambda = 0. \tag{4.1}$$

The conditions for which the two null hypersurfaces $u = 0$ and $v = 0$ are free of matter are

$$U_{,u}(u = 0, v) = 0 = U_{,v}(u, v = 0). \tag{4.2}$$

For the special gauge of Eq. (3.40), these conditions together with Eq. (3.38) imply

$$n, m > \frac{1}{2}. \tag{4.3}$$

In terms of η and μ, the Einstein vacuum equations (Chandrasekhar and Ferrari, 1984) can be written in the form,

$$\text{Re}(Z)\{[(1 - \eta^2)\, Z_{,\eta}]_{,\eta} - [(1 - \mu^2)\, Z_{,\mu}]_{,\mu}\}$$
$$= (1 - \eta^2)\, (Z_{,\eta})^2 - (1 - \mu^2)\, (Z_{,\mu})^2, \tag{4.4}$$

and

$$\frac{\mu}{1 - \mu^2}\Gamma_{,\eta} + \frac{\eta}{1 - \eta^2}\Gamma_{,\mu} = -\frac{2}{(Z + \overline{Z})^2}\,(Z_{,\eta}\overline{Z}_{,\mu} + \overline{Z}_{,\eta}Z_{,\mu}), \tag{4.5a}$$

$$2\eta\Gamma_{,\eta} + 2\mu\Gamma_{,\mu} = \frac{3}{1 - \eta^2} + \frac{1}{1 - \mu^2}$$
$$- \frac{4}{(Z + \overline{Z})^2}[(1 - \eta^2)\,|Z_{,\eta}|^2 + (1 - \mu^2)\,|Z_{,\mu}|^2], \tag{4.5b}$$

where

$$f \equiv \frac{(1 - \eta^2)^{1/2}}{\eta^2 - \mu^2}e^{\Gamma}. \tag{4.6}$$

Equation (4.4) is equivalent to the Ernst equation (3.43) with the Ernst potential E defined by Eq. (3.44). In terms of the Ernst potential E, Eqs. (4.5a)–(4.5b) can be written as

$$\frac{\mu}{1 - \mu^2}\Gamma_{,\eta} + \frac{\eta}{1 - \eta^2}\Gamma_{,\mu} = -\frac{2}{(1 - |E|^2)^2}\,(E_{,\eta}\,\overline{E}_{,\mu} + \overline{E}_{,\eta}\,E_{,\mu}), \tag{4.7a}$$

$$2\eta\Gamma_{,\eta} + 2\mu\Gamma_{,\mu} = \frac{3}{1 - \eta^2} + \frac{1}{1 - \mu^2}$$
$$- \frac{4}{(1 - |E|^2)^2}[(1 - \eta^2)\,|E_{,\eta}|^2 + (1 - \mu^2)\,|E_{,\mu}|^2]. \tag{4.7b}$$

It can be shown that the above equations are equivalent to the ones given by Eqs. (3.94)–(3.97).

On the other hand, Chandrasekhar and Ferrari (1984) showed that, as we did in the stationary axisymmetric case, if we introduce a potential Q for the function q_2 in the manner,

$$Q_{,\eta} = \frac{1 - \mu^2}{\chi^2} q_{2,\mu}, \quad Q_{,\mu} = \frac{1 - \eta^2}{\chi^2} q_{2,\eta}, \tag{4.8}$$

then the complex function Z^+ defined by

$$Z^+ \equiv P + iQ \tag{4.9}$$

satisfies the same equation (4.4), where

$$P \equiv \left(1 - \eta^2\right)^{1/2} \left(1 - \mu^2\right)^{1/2} \chi^{-1}. \tag{4.10}$$

Thus, in contrast to the stationary axisymmetric case, we can obtain the Ernst equation either directly from the metric coefficients χ and q_2 or from the potential Q. When the approaching waves have aligned constant polarizations, we can globally set

$$q_2 = 0 \quad (\text{or } W = 0). \tag{4.11}$$

Then, the Einstein vacuum equations of Eqs. (4.4), (4.5a) and (4.5a) reduce to

$$\chi\{[(1 - \eta^2)\chi_{,\eta}]_{,\eta} - [(1 - \mu^2)\chi_{,\mu}]_{,\mu}\}$$
$$= \left(1 - \eta^2\right) (\chi_{,\eta})^2 - \left(1 - \mu^2\right) (\chi_{,\mu})^2, \tag{4.12}$$

and

$$\frac{\mu}{1 - \mu^2}\Gamma_{,\eta} + \frac{\eta}{1 - \eta^2}\Gamma_{,\mu} = -\frac{\chi_{,\eta}\chi_{,\mu}}{\chi^2}, \tag{4.13a}$$

$$2\eta\Gamma_{,\eta} + 2\mu\Gamma_{,\mu} = \frac{3}{1 - \eta^2} + \frac{1}{1 - \mu^2} - \frac{\left(1 - \eta^2\right)\chi_{,\eta}^2 + \left(1 - \mu^2\right)\chi_{,\mu}^2}{\chi^2}. \tag{4.13b}$$

In terms of V [see Eq. (3.44)], Eq. (4.12) takes the form

$$\left[(1 - \eta^2)\,V_{,\eta}\right]_{,\eta} - \left[(1 - \mu^2)\,V_{,\mu}\right]_{,\mu} = 0. \tag{4.14}$$

An obvious solution of Eq. (4.14) is given by (Stoyanov, 1979)

$$V_1 = a\ln\left[\left(1 - \eta^2\right)\left(1 - \mu^2\right)\right], \tag{4.15}$$

where a is an arbitrary constant.

Separation of variables, on the other hand, leads to another class of solutions (Chandrasekhar and Xanthopoulos, 1986a; Griffiths, 1987; Tsoubelis and Wang, 1989),

$$V_2 = \sum_{n=0}^{\infty} [A_n P_n(\eta) P_n(\mu) + B_n P_n(\eta) Q_n(\mu) + C_n Q_n(\eta) P_n(\mu)$$
$$+ D_n Q_n(\eta) Q_n(\mu)], \tag{4.16}$$

where A_n, B_n, C_n and D_n are arbitrary constants, and $P_n(\eta)$ and $Q_n(\eta)$ are the Legendre functions of the first and second kind, respectively.

Note that Feinstein and Ibañez (1989) gave another expression for the general solutions of V. However, a complete solution of the Einstein field equations is obtained only after the integration of Eqs. (4.13a) and (4.13b) for the function Γ is carried out explicitly. But, for the general solutions,

$$V = V_1 + V_2, \tag{4.17}$$

such an integration has not been completely solved yet, and only for some particular choice of V, the corresponding solutions of Γ have been found (Griffiths, 1987; Tsoubelis and Wang, 1989; 1992; Li, 1989; Wang, 1991a).

In the following, let us consider the solutions (Tsoubelis and Wang, 1989),

$$V = a \ln \left[\left(1 - \eta^2\right) \left(1 - \mu^2\right) \right] - 2\delta_1 Q_0(\eta) P_0(\mu) - 2\delta_2 P_0(\eta) Q_0(\mu)$$
$$= a \ln \left[\left(1 - \eta^2\right) \left(1 - \mu^2\right) \right] + \delta_1 \ln \frac{1 - \eta}{1 + \eta} + \delta_2 \ln \frac{1 - \mu}{1 + \mu}, \tag{4.18}$$

where δ_1 and δ_2 are two constants related to the parameters n and m introduced in Section 3.3.

In order to complete the integration of Eqs. (4.13a) and (4.13b), we introduce the quantity Σ by

$$\Gamma = \ln \left(\frac{\eta^2 - \mu^2}{\left(1 - \eta^2\right)^{3/4} \left(1 - \mu^2\right)^{1/4}} \right) + \Sigma. \tag{4.19}$$

Substituting the above expression into Eqs. (4.13a) and (4.13b), we obtain

$$\frac{\mu}{1 - \mu^2} \Sigma_{,\eta} + \frac{\eta}{1 - \eta^2} \Sigma_{,\mu} = -V_{,\eta} V_{,\mu}, \tag{4.20a}$$

$$2\eta \Sigma_{,\eta} + 2\mu \Sigma_{,\mu} = -\left(1 - \eta^2\right) \left(V_{,\eta}\right)^2 - \left(1 - \mu^2\right) \left(V_{,\mu}\right)^2. \tag{4.20b}$$

It can be shown that for the function V given by Eq. (4.18), the system of Eqs. (4.20a) and (4.20b) has the following solution (Tsoubelis and Wang, 1989):

$$\Sigma = -\delta_+^2 \ln(\eta + \mu) - \delta_-^2 \ln(\eta - \mu) + (a + \delta_1)^2 \ln(1 - \eta)$$
$$+ (a - \delta_1)^2 \ln(1 + \eta) + (a + \delta_2)^2 \ln(1 - \mu)$$
$$+ (a - \delta_2)^2 \ln(1 + \mu) + \ln C_0, \tag{4.21}$$

where C_0 is an arbitrary constant, and $\delta_\pm$ are defined by

$$\delta_\pm \equiv \delta_1 \pm \delta_2. \tag{4.22}$$

To summarize the above results, we obtain the following solutions:

$$f = \frac{C_0}{(\eta + \mu)^{\delta_+^2}(\eta - \mu)^{\delta_-^2}}(1 - \eta)^{b_1}(1 - \mu)^{b_2}(1 + \eta)^{c_1}(1 + \eta)^{c_2},$$

$$e^{-U} = \left[(1 - \eta^2)(1 - \mu^2)\right]^{1/2},$$

$$e^V = \left[(1 - \eta^2)(1 - \mu^2)\right]^a \left(\frac{1 - \eta}{1 + \eta}\right)^{\delta_1}\left(\frac{1 - \mu}{1 + \mu}\right)^{\delta_2}, \tag{4.23}$$

$$W = 0,$$

where

$$b_A = (a + \delta_A)^2 - \frac{1}{4}, \quad c_A = (a - \delta_A)^2 - \frac{1}{4}, \quad (A = 1, 2). \tag{4.24}$$

The solutions of Eq. (4.23) include most of the known diagonal ($W = 0$) solutions. For example, when $a = 0$, we obtain Szekeres' family of the colliding plane gravitational wave solutions (Szekeres, 1972), which includes the Szekeres solution (Szekeres, 1970), in which we have $n = m = 2$, and the Khan–Penrose solution (Khan and Penrose, 1971), in which $n = m = 1$, where n and m are defined below in Eq. (4.28). For any given a, when $\delta_1 = 1$ and $\delta_2 = 0$, we obtain the Ferrari–Ibañez solutions (Ferrari and Ibañez, 1987a).

In terms of t and z, the function V given by Eq. (4.23) can be written as

$$V = 2a \ln t - \delta_+ \ln\left[\frac{1 + z + \sqrt{(-1 - z)^2 - t^2}}{t}\right]$$
$$- \delta_- \ln\left[\frac{1 - z + \sqrt{(1 - z)^2 - t^2}}{t}\right]. \tag{4.25}$$

The last two terms in the right-hand side of Eq. (4.25) correspond to the soliton structure, while the first term can be thought of as producing a

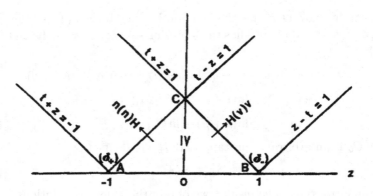

Fig. 4.1.   The soliton structure for the solutions given by Eq. (4.25).

homogeneous expansion and is mainly "responsible" for the development of the spacetime curvature singularities as $t \to 0^+$ (Feinstein and Ibañez, 1989). When $\delta_- = 1$, there is a single pole located at the point $z = 1$ [see Fig. 4.1]. When $\delta_- = 2$, there is a double pole located at that point, and when δ_- takes continuous values, there are "generalized" poles located at that point (Verdaguer, 1984; Feinstein and Charach, 1986; Oliver and Verdaguer, 1989; Carot and Verdaguer, 1989). Similarly, the δ_+ term in Eq. (4.25) corresponds to the poles located at the point $z = -1$. The contribution of the solitonic terms to the divergence of the curvature as $t \to 0^+$ is similar to that of the homogenous term.

The expression of Eq. (4.25) makes the solutions valid in the region outside of the two light-cones

$$(1 - z)^2 - t^2 = 0, \quad (-1 - z)^2 - t^2 = 0, \tag{4.26}$$

and Eq. (3.49) further restricts the region of validity of the solutions to that between the two light-cones [Region IV in Fig. 4.1].

Thus, the Khan–Penrose substitutions of Eq. (3.11) actually extend these solutions defined in the triangle ABC beyond the lines AC and BC to Regions I–III, as shown in Fig. 4.1.

On the other hand, combining Eqs. (3.42) and (4.23) we find

$$e^{-M} = 8nm u^{2n-1} v^{2m-1} f$$
$$= \frac{u^{n(2-\delta_+^2)-1} v^{m(2-\delta_-^2)-1}}{X^{\delta_+^2} Y^{\delta_-^2}}$$
$$\times (1 - \eta)^{b_1} (1 - \mu)^{b_2} (1 + \eta)^{c_1} (1 + \mu)^{c_2}. \tag{4.27}$$

In writing Eq. (4.27) we have set $8nm2^{-(\delta_+ +\delta_-)}C_0 = 1$ by using the arbitrariness of C_0. Equation (4.27) shows that the application of the Khan–Penrose substitutions Eq. (3.11) will result in singularity at $u = 0$ or $v = 0$, unless the parameters n and m are chosen so that

$$n = \frac{1}{2 - \delta_+^2}, \quad m = \frac{1}{2 - \delta_-^2}. \tag{4.28}$$

Equations (4.3) and (4.28) are the restrictions on the choice of the soliton parameters δ_1 and δ_2. The solutions which do not satisfy these conditions are not acceptable for the collision of purely plane gravitational waves. However, further considerations from the point of view of physics (Tsoubelis and Wang, 1989) show that the conditions of Eq. (4.3) are too weak and that for $1/2 < m, n < 1$ the extension obtained above cannot guarantee the integrability of the Weyl scalars corresponding to the waves under collision (Szekeres, 1972), as it will become clear in the following. Therefore, we impose the conditions

$$n, m \geq 1, \tag{4.29}$$

so as to make the Weyl scalars integrable (Szekeres, 1972; Tsoubelis and Wang, 1989). Then, the extended solutions in the pre-collision regions are given as follows.

In Region I, where $u, v < 0$, we have [see Eq. (3.22)],

$$M = U = V = 0, \tag{4.30}$$

and the corresponding metric takes the form

$$ds^2 = 2dudv - \left(dx^2\right)^2 - \left(dx^3\right)^2, \tag{4.31}$$

which is flat.

In Region II, where $u < 0$, and $0 < v < 1$, we have

$$
\begin{aligned}
e^{-M} &= (1 + v^m)^{b_2 + c_1 + (1-n)/2n - 1/2} (1 - v^m)^{b_1 + c_2 + (1-n)/2n - 1/2}, \\
e^{-U} &= 1 - v^{2m}, \\
e^{V} &= (1 + v^m)^{2a - \delta_-} (1 - v^m)^{2a + \delta_-}, \\
W &= 0.
\end{aligned}
\tag{4.32}
$$

Finally, in Region III, where $0 < u < 1$, and $v < 0$, we have

$$e^{-M} = (1 + u^n)^{c_1 + c_2 + (1-m)/2m - 1/2} (1 - u^n)^{b_1 + b_2 + (1-m)/2m - 1/2},$$

$$e^{-U} = 1 - u^{2n},$$
$$eV = (1 + u^n)^{2a - \delta_+} (1 - u^n)^{2a + \delta_+},$$
$$W = 0.$$
(4.33)

Substituting Eq. (4.32) into Eqs. (3.7a)–(3.7c), we find that the only non-vanishing Weyl scalar is Ψ_0, given by

$$\Psi_0^{II}(v) = -e^{2U} \{ 2am^2 \left(1 - 4a^2 \right) v^{4m-2} - 12a^2 m^2 \delta_- v^{3m-2}$$
$$- 6am(2m - 1)v^{2m-2} - m(m - 1)\delta_- v^{m-2} \}.$$
(4.34)

Similarly, ψ_4^{III} is the only non-vanishing Weyl scalar in Region III, and is given by

$$\Psi_4^{III}(u) = -e^{2U} \{ 2an^2(1 - 4a^2)u^{4n-2} - 12a^2 n^2 \delta_+ u^{3n-2}$$
$$- 6an(2n - 1)u^{2n-2} - n(n - 1)\delta_+ u^{n-2} \}.$$
(4.35)

In Region IV, all three of the Weyl scalars Ψ_0, Ψ_2 and Ψ_4 do not vanish, but the corresponding expressions are very complicated, so we shall not present them explicitly here. Anyhow, as far as the behavior of the Weyl scalars across the hypersurfaces $u = 0$ and $v = 0$ is concerned, we only need to know their limits as $u \to 0^+$ and $v \to 0^+$, which are given as follows.

When $u \to 0^+$, we find

$$\Psi_0^{IV}(0, v) \to \Psi_0^{II}(v),$$
$$\Psi_2^{IV}(0, v) \to \frac{nm\delta_+}{X^3} u^{n-1} v^{m-1} (2av^m + \delta_-),$$
$$\Psi_4^{IV}(0, v) \to \frac{n}{X^2} u^{n-2} [(n - 1)\delta_+ X + 6a(2n - 1)u^n$$
$$+ 3(2n - 1)\delta_- v^m u^n],$$
(4.36)

and when $v \to 0^+$, we have

$$\Psi_0^{IV}(u, 0) \to \frac{m}{Y^2} v^{m-2} [(m - 1)\delta_- Y + 6a(2m - 1)v^m$$
$$+ 3(2m - 1)\delta_+ v^m u^n],$$
$$\Psi_2^{IV}(u, 0) \to \frac{nm\delta_-}{Y^3} u^{n-1} v^{m-1} (2au^n + \delta_+),$$
$$\Psi_4^{IV}(u, 0) \to \Psi_4^{II}(u).$$
(4.37)

Combining the above results with Eqs. (3.24) and (3.26), it can be seen that the Weyl scalars behave as follows when we cross the null hypersurface separating region A from region B ($A \to B$):

(i) IV → II:

$$\Psi_0 \text{ is continuous,}$$
$$\Psi_2 = \Psi_2^{IV} H(u), \tag{4.38}$$
$$\Psi_4 = \Psi_4^{IV} H(u) + \frac{n\delta_+}{X} u^{n-1}\delta(u),$$

(ii) IV → III:

$$\Psi_0 = \Psi_0^{IV} H(v) + \frac{m\delta_-}{Y} v^{m-1}\delta(v),$$
$$\Psi_2 = \Psi_2^{IV} H(v), \tag{4.39}$$
$$\psi_4 \text{ is continuous,}$$

(iii) II → I:

$$\Psi_0 = \Psi_0^{II} H(v) + m\delta_- v^{m-1}\delta(v),$$
$$\Psi_2, \Psi_4 \text{ are continuous,} \tag{4.40}$$

(iv) III → I:

$$\Psi_2, \Psi_4 \text{ are continuous,}$$
$$\Psi_4 = \Psi_4^{III} H(u) + n\delta_+ u^{n-1}\delta(u). \tag{4.41}$$

Equations (4.34) and (4.40) show clearly that the parameter m determines the type of the wave incident in Region II. Particularly, by observing the behavior of Ψ_0 across the hypersurface $v = 0$ in the direction $II \to I$, we can distinguish the following cases.

(a) $m = 1$. In this case, we have

$$\Psi_0 = 6aH(v) + \delta_-\delta(v). \tag{4.42}$$

Therefore, when $a \neq 0$, the gravitational wave incident in Region II has the form of an impulsive + shock wave. When $a = 0$, only the impulsive part remains.

(b) $1 < m < 2$. Then, we have

$$\Psi_0 = m(m-1)\delta_- v^{m-2}H(v), \tag{4.43}$$

which means that the incoming gravitational wave has an unbounded wave-front of the form v^γ, with $\gamma \in (0, 1)$.

(c) $m = 2$. Then, we have

$$\Psi_0 = 2\delta_- H(v), \tag{4.44}$$

which corresponds to a shock wave.

(d) $m > 2$. Then, we find that Ψ_0 is continuous across this hypersurface. Thus, the wavefront in this case is smooth.

Replacing v by u and m by n (and, therefore, δ_- by δ_+) in Eqs. (4.42)–(4.44), we obtain the type of the wave incident in Region III. It is then obvious that, by properly choosing the values of the parameters a, m, and n, we can have a variety of situations, for example, an impulsive wave collides with a shock wave, or an impulsive + shock wave, or a wave with smooth wavefront, and so on.

Having finished the studies of the behavior of the Weyl scalars across the hypersurfaces $u = 0$ and $v = 0$, we now turn to study the behavior of these scalars near the hypersurface $t = 0$ (or equivalently, $\eta = 1$, or $u^{2n} + v^{2m} = 1$), since this will reveal the singular properties of the spacetimes as the result of the collision of two pure plane gravitational waves. To this purpose, let us first return to the expression (3.52), from which we find that the determinant of the metric is given by

$$\det\left[g_{\mu\nu}(\eta, \mu)\right] = -\left(1 - \eta^2\right)e^{2\Gamma}$$

$$= -C_0^2 \frac{(1-\eta)^{2b_1}(1-\mu)^{2b_2}(1+\eta)^{2c_1}(1+\mu)^{2c_2}}{(\eta+\mu)^{2\delta_+^2-2}(\eta-\mu)^{2\delta_-^2-2}}. \quad (4.45)$$

Thus, as $t \to 0^+$ or $\eta \to 1^-$, the metric becomes singular, unless

$$b_1 = 0, \quad (4.46)$$

which is equivalent to

$$(a + \delta_1)^2 = \frac{1}{4}, \quad (4.47)$$

as one can see from Eq. (4.24).

Detailed calculations show that Eq. (4.47) is also the sufficient condition for which the Weyl scalars remain bounded as $t \to 0^+$. Specifically, when Eq. (4.47) holds, the non-vanishing Weyl scalars have the following limits:

$$\Psi_2^{IV} \to mn \, 2^{-(b_2+c_1+c_2+2)} u^{-2(nc_2+1)} v^{-2(mb_2+1)}$$

$$\times \left[\frac{1}{4} - (a - \delta_2)^2 - 4au^{2n}(\delta_2 + \delta_1 v^{2m})\right],$$

$$\Psi_0^{IV} \to -\frac{3mu^{1-2n}v^{2m-1}}{2n(a+\delta_1)}\Psi_2^{IV}, \quad (4.48)$$

$$\Psi_4^{IV} \to -\frac{3nu^{2n-1}v^{1-2m}}{2m(a+\delta_1)}\Psi_2^{IV},$$

as $t \rightarrow 0^+$. Therefore, when Eq. (4.47) holds, no curvature singularities develop on the "focusing hypersurface" $t = 0$, and instead, a Cauchy horizon is developed. Thus, in this case the metric is extendible across the hypersurface $t = 0$. However, as pointed out by Yurtsever (1988a, 1989), the geometry of the spacetime beyond this hypersurface is not uniquely determined by the initial data posed on the two intersecting characteristic hypersurfaces $u = 0$ and $v = 0$. The spacetime can be smoothly extended across the Cauchy horizon in infinitely different ways.

In addition, Griffiths (2005) rigorously proved that such Cauchy horizons are not stable, and will be turned into spacetime singularities by general bounded perturbations of the initial waves.

From Eq. (4.48) it can be shown that

$$9 \left(\Psi_2^{IV} \right)^2 \rightarrow \Psi_0^{IV} \Psi_4^{IV}, \tag{4.49}$$

as $t \rightarrow 0^+$. Then, following the theorem given by Chandrasekhar and Xanthopoulos (1986a) that, when the relation

$$9 \left(\Psi_2^{IV} \right)^2 = \Psi_0^{IV} \Psi_4^{IV}$$

holds, the corresponding solutions are Petrov type D, we find from Eq. (4.49) that the above solutions become Petrov type D as $t \rightarrow 0^+$. It is worth noting in this regard that the extendible colliding plane gravitational wave models obtained by Chandrasekhar and Xanthopoulos (1986a), and Ferrari and Ibañez (1988) are also Petrov type D.

On the other hand, if we follow Yurtsever's approach (1988a, 1989), we find that the solutions given by Eqs. (4.23) and (4.24) have the limit,

$$ds^2 \approx \varepsilon_0^{(0)} d\tau^2 - \varepsilon_1^{(0)} \tau^{2p_1} dz^2 - \varepsilon_2^{(0)} \tau^{2p_2} \left(dx^2 \right)^2 - \varepsilon_3^{(0)} \tau^{2p_3} \left(dx^3 \right)^2, \tag{4.50}$$

as $t \rightarrow 0^+$, where

$$\tau = t^{\alpha^2 + 3/4}, \quad \alpha = a + \delta_1, \tag{4.51}$$

and

$$p_1 \equiv \frac{\alpha^2 - \frac{1}{4}}{\alpha^2 + \frac{3}{4}}, \quad p_2 \equiv -\frac{\alpha - \frac{1}{2}}{\alpha^2 + \frac{3}{4}}, \quad p_3 \equiv \frac{\alpha + \frac{1}{2}}{\alpha^2 + \frac{3}{4}}, \tag{4.52}$$

and $\varepsilon_\mu^{(0)}$ are functions of z only. It is easy to show that the p_k defined by Eq. (4.52) satisfy the Kasner relations Eq. (3.92), but now the p_k

are independent of the coordinate z. Note that the condition (3.93) is equivalent to

$$\alpha = \pm\frac{1}{2}, \tag{4.53}$$

which are exactly the conditions given by Eq. (4.47).

Equations (4.50)–(4.53) show clearly that the homogeneous expansion parameter a, together with the solitonic parameter δ_1, completely determines the singularity behavior of the solutions on the focusing hypersurface $t = 0$. Specifically, when $\alpha < -1/2$, p_1 and p_2 are greater than zero, while p_3 is less than zero. Thus, the corresponding singularities are astigmatic (Yurtsever, 1988a, 1989). That is, the physical three-values get squashed in the z and x^2 directions and stretched in the x^3 direction. When $\alpha \to \infty$, $p_1 \to 1$, $p_2 \to 0^+$, and $p_3 \to 0^-$. Thus, in this case the solutions approach the degenerate Kasner solution. When $\alpha = -1/2$, as shown above, the corresponding solutions are free of curvature singularities on the focusing hypersurface $t = 0$. When $-1/2 < a \le 0$, then we have $p_1 < 0$. Consequently, the corresponding singularities on $t = 0$ become anastigmatic, which means that the physical three-values now become stretched in the z direction and squashed in the x^2 and x^3 directions. Because of the symmetric dependence of the exponents p_k on a, it is easy to see that the same results can be obtained for $\alpha \ge 0$, if x^2 and x^3 are exchanged (and, therefore, p_2 and p_3 are exchanged).

Finally, we note that other solutions, which represent the collisions of purely collinearly polarized gravitational plane waves, were also found by several authors (Ferrari and Ibañez, 1987b; Griffiths, 1987; Yurtsever, 1988b; Li, 1989; Tomita, 1998). However, from the point of view of physics, these solutions share similar physical properties of the ones just considered above.

4.2. Collisions of Collinearly and Non-collinearly Polarized Gravitational Plane Waves

Solutions which represent the collision of non-collinearly polarized gravitational plane waves were first found by Nutku and Halil (1977) in generalizing the Khan–Penrose solution (Khan and Penrose, 1971) to the non-collinear case. The Nutku–Halil solution represents the collision of two constantly but not collinearly polarized impulsive gravitational plane waves. Later on, several other solutions were found (Chandrasekhar and Xanthopoulos, 1986a; Ferrari Ibañez and Bruni, 1987a; Ernst, Garcia and Hauser, 1987a; Halilsoy, 1988a,1988b; Tsoubelis and Wang, 1992). Except for the Tsoubelis–Wang solutions, which represent the collision of a collinearly with a non-collinearly polarized gravitational plane wave, all other solutions mentioned above represent the collisions of two non-collinearly polarized gravitational plane waves.

In this section, we shall present a four-parameter class of solutions obtained by introducing one soliton into the diagonal solutions given by Eqs. (2.31) and (2.32), which in general represents the collision of two plane gravitational waves, one of which is collinearly polarized, while the other is non-collinearly polarized (Tsoubelis and Wang, 1992).

4.2.1. One-soliton solutions

As mentioned in Section 3.6, the crucial point in using the inverse scattering method of Belinsky and Zakharov is to integrate the system (3.115) for a given (diagonal) seed solution. But, this in general is not an easy task. Fortunately, for the seed solutions given by Eqs. (4.23) and (4.24), after a simple but tedious integration, the following solution can be obtained:

$$\Sigma^{(0)}(\lambda, \eta, \mu) = \frac{(2\lambda w)^{a+\delta_1}}{[(1+\eta)(1+\mu) + \lambda]^{\delta_+}[(1+\eta)(1-\mu) - \lambda]^{\delta_-}}. \qquad (4.54)$$

Then, the remaining task is no more than some algebraic calculations. By choosing the pole trajectory as

$$\mu_1 \equiv 1 - z - \sqrt{(1-z)^2 - t^2} = (1-\eta)(1+\mu), \qquad (4.55)$$

which is equivalent to introducing one soliton (or pole) at the point $z = 1$, we find that the one-soliton solutions are given by

$$\chi = \left[(1-\eta^2)(1-\mu^2)\right]^{1/2} \frac{A}{B} \chi^{(0)},$$

$$q_2 = -2q \frac{(\eta - \mu)^{2\delta_- +1}}{B} [\chi^{(0)}]^2, \qquad (4.56)$$

$$f = C_{ph}^{(1)} \frac{(1-\eta)^{b_1'}(1-\mu)^{b_2'}(1+\eta)^{c_1'}(1+\mu)^{c_2'}}{(\eta + \mu)^{\delta_+^2}(\eta - \mu)^{(\delta_- +1)^2}} A,$$

where

$$A \equiv (1-\eta)^{2(a+\delta_1)}(1+\mu)^{2(a-\delta_2)} + q^2(\eta - \mu)^{4\delta}$$
$$- (1+\eta)^{2(a-\delta_1)}(1-\mu)^{2(a+\delta_2)},$$

$$B \equiv (1-\eta)^{2(\alpha+\delta_1)}(1+\mu)^{2(a-\delta_2)}(1+\eta)(1-\mu) \qquad (4.57)$$
$$+ q^2(\eta - \mu)^{4\delta} - (1+\eta)^{2(a-\delta_1)}(1-\mu)^{2(a+\delta_2)}(1-\eta)(1+\mu),$$

with

$$b'_A \equiv (a + \delta_A)(a + \delta_A - 1), \quad c'_A \equiv (a - \delta_A)(a - \delta_A - 1), \qquad (4.58)$$

where $q \left(\equiv 4^{\delta_1 - a} Q_1 \right)$ is another arbitrary constant.

Note that when $\eta \to \mu$ the functions A and B defined by Eq. (4.57) become unbounded unless $\delta_- \geq 0$. Therefore, for the extension of the above solutions beyond the hypersurfaces $\eta = \pm \mu$ (equivalent to the $u = 0$ and $v = 0$ hypersurfaces), we restrict the solutions of Eqs. (4.56)–(4.58) only to the cases in which $\delta_- \geq 0$. For $\delta_- < 0$, we write these solutions in the form

$$\chi = [(1 - \eta^2)(1 - \eta^2)]^{1/2} \frac{A'}{B'} \chi^{(0)},$$

$$q_2 = -2q \frac{(\eta - \mu)^{-2\delta_- + 1}}{B'} [\chi^{(0)}]^2, \qquad (4.59)$$

$$f = C_{ph}^{(1)} \frac{(1 - \eta)^{b'_1}(1 - \mu)^{b'_2}(1 + \eta)^{c'_1}(1 + \mu)^{c'_2}}{(\eta + \mu)^{\delta_+^2}(\eta - \mu)^{(\delta_- - 1)^2}} A',$$

where

$$A' \equiv (\eta - \mu)^{-4\delta_-} A, \quad B' \equiv (\eta - \mu)^{-4\delta_-} B, \qquad (4.60)$$

and A and B are defined by Eq. (4.57).

When we introduce one pole at the point $z = -1$ [see Fig. 4.1], we find that the corresponding solutions are given by Eqs. (4.56)–(4.58) but replacing μ by $-\mu$ and δ_2 by $-\delta_2$. Without loss of generality, in the following we shall consider only the solutions in which $\delta_- \geq 0$.

The considerations given in Section 4.1 show that the proper choice of the parameter n and m for the solutions given by Eqs. (4.56)–(4.58) is

$$n = (2 - \delta_+^2)^{-1}, \quad m = [2 - (\delta_- + 1)^2]^{-1}. \qquad (4.61)$$

The restrictions of Eq. (4.29) now imply that

$$1 \leq \delta_+^2 < 2, \quad 0 \leq \delta_- < \sqrt{2} - 1. \qquad (4.62)$$

Inserting the Khan–Penrose substitutions Eq. (3.11) into the above solutions and considering Eq. (3.42), we find that the extended solutions in the pre-collision regions are given as follows.

In Region I, the metric takes the form of Eq. (4.31), which means that the spacetime is flat.

In Region II, the solutions are given by

$$\chi = (1 - v^m)^{2a+\delta_- +1} (1 + v^m)^{2a-\delta_- +1} \frac{A}{B},$$

$$q_2 = -4 \left(\frac{q'}{B}\right) v^{(2\delta_- +1)m} (1 - v^m)^{2(2a+\delta_-)} (1 + v^m)^{2(2a-\delta_-)},$$

$$e^{-M} = A (1 - v^m)^{2a(a+\delta_- -1)+\delta_-(\delta_- -2)/2}$$
$$\times (1 + v^m)^{2a(a-\delta_- -1)+\delta_-(\delta_- +2)/2}, \tag{4.63}$$

where

$$A = (1 - v^m)^{2(2a+\delta_-)} + q'^2 v^{4\delta_- -m} (1 + v^m)^{2(2a-\delta_-)},$$

$$B = (1 - v^m)^{2(2a+\delta_-)}(1 + v^m)^2 + q'^2 v^{4\delta_- -m}(1 + v^m)^{2(2a-\delta_-)}(1 - v^m)^2,$$

$$q' \equiv 4^{\delta_-} q. \tag{4.64}$$

Note that when writing Eq. (4.64) we had set $8mn2^{-\delta_+^2 -(1-\delta_-)^2} C^{(1)}_{ph} = 1$.

In Region III, we have

$$\chi = (1 - u^n)^{2a+\delta_+} (1 + u^n)^{2a-\delta_+},$$

$$q_2 = 0, \tag{4.65}$$

$$e^{-M} = (1 - u^n)^{2a(a+\delta_+)+(\delta_+^2 -1)/2}(1 + u^n)^{2a(a-\delta_+)+(\delta_+^2 -1)/2}.$$

Equations (4.63) and (4.65) show that the extended solutions represent the collisions of two plane gravitational waves. The one incident in Region II is in general non-collinearly polarized, while the one in Region III is collinearly polarized. It is interesting to note that the solutions in Region III are exactly the ones given by Eq. (4.33) in the collinear case. Thus, the types that the gravitational plane wave incident in Region III may have are the same ones as discussed in Section 4.1. The types that the wave incident in Region II may have are given as follows.

(i) $m = 1$. Then, we have

$$\Psi_0^{II\to I} = \frac{6a}{(1 + q^2)^2}[(1 - 6q^2 + q^4) - i4q(1 - q^2)]H(v)$$

$$+ \frac{1}{1 + q^2}[(1 - q^2) - i2q]\delta(v). \tag{4.66}$$

Thus, as it is in the collinear case, the incoming gravitational plane wave in Region II is the type of impulsive + shock. When $a = 0$, only the impulsive part remains.

(ii) $1 < m < 2$. In this case, we have

$$\Psi_0^{II\to I} \to m(m-1)(1+\delta_-)v^{m-2}$$
$$- i2m(1+2\delta_-)(2m\delta_- + m - 1)v^{m(2\delta_-+1)-2}, \qquad (4.67)$$

which means that the incoming wave has unbounded wavefront when across the hypersurface $v = 0$.

(iii) $m = 2$. Then, we have

$$\Psi_0^{II\to I} = 2(1+\delta_-) H(v), \qquad (4.68)$$

which is a shock wave only.

(iv) $m > 2$. Then, we find that Ψ_0 is continuous across the hypersurface $v = 0$. So, the incoming gravitational wave has smooth wavefront in this case.

Combining the above results with the ones obtained in the collinear case, we find that the values of the parameters n and m uniquely determine the type of the incoming gravitational plane waves, no matter whether the waves are constantly (collinearly) polarized or variably (non-collinearly) polarized. The freedom for the choice of the values of the parameters n and m leads the above solutions to represent a variety of models, for example, an impulsive plane gravitational wave collides with a shock wave, or a shock + impulsive wave, or a wave with a smooth wavefront, etc.

4.2.2. The nature of singularities formed after collision

To study the nature of the singularities of the solutions presented in the last subsection on the hypersurface $t = 0$, we begin with their diagonal limit. Setting $q = 0$ in Eqs. (4.56)–(4.58), we find that the corresponding solutions are given by

$$\chi = \left[(1-\eta^2)(1-\mu^2)\right]^a \left[\frac{1-\eta}{1+\eta}\right]^{\delta_1} \left[\frac{1-\mu}{1+\mu}\right]^{\delta_2},$$

$$q_2 = 0, \qquad (4.69)$$

$$e^{-M} = \frac{(1-\eta)^{b_1''}(1-\mu)^{b_2''}(1+\eta)^{c_1''}(1+\eta)^{c_2''}}{X^{\delta'+{}^2} + Y^{\delta'-{}^2}},$$

where

$$b_A'' = (a+\delta_A')^2 - \frac{1}{4}, \quad c_A'' = (a-\delta_A')^2 - \frac{1}{4},$$
$$\delta_1' = \delta_1 + \frac{1}{2}, \quad \delta_2' = \delta_2 - \frac{1}{2}, \quad \delta'\pm = \delta_1' \pm \delta_2'. \qquad (4.70)$$

But, after replacing δ_1 and δ_2 by δ_1' and δ_2', respectively, these are the seed solutions given by Eqs. (4.23) and (4.24). Thus, the asymptotic behavior of the solutions given by Eq. (4.69) is described by Eqs. (4.50)–(4.52) but with α being replaced by $\alpha' \equiv a + \delta_1' = \alpha + 1/2$. Specifically, corresponding to Eq. (4.53), we now have

$$\alpha = 0, \ -1. \qquad (4.71)$$

In the non-collinear case, the solutions given by Eqs. (4.56)–(4.58) have the following limit as $t \to 0^+$.

(a) $\alpha < -1/2$. In this case, it can be shown that the above solutions have the same limit as the corresponding diagonal solutions as $t \to 0^+$. Specifically, the solutions with $\alpha = -1$ are free of spacetime curvature singularities on the hypersurface $t = 0$ in both of the collinear and non-collinear cases.

(b) $-1/2 \leq \alpha < 0$. In this case, the solutions have the limit

$$ds^2 = \varepsilon_0^2 d\tau^2 - \varepsilon_1^2 \tau^{2p_1} dz^2 - \varepsilon_2^2 \tau^{2p_2} (dX_{(z)}^2)^2 - \varepsilon_3^2 \tau^{2p_3} (dX_{(z)}^3)^2, \qquad (4.72)$$

as $t \to 0^+$, where τ and p_k are given by Eqs. (4.51) and (4.52) with α being replaced by $\alpha' = \alpha + 1/2$. Here ε_i^2's are functions of z only, and $X_{(z)}^2$ and $X_{(z)}^3$ are z-dependent linear combinations of x^2 and x^3 at each fixed point of z. Equation (4.72) shows that, relative to the corresponding diagonal solutions, the Kasner exponents p_k are not changed. Thus, in the both collinear and non-collinear cases, the nature of singularities on the hypersurface $t = 0$ is the same, and is anastigmatic. However, the Kasner axes along which the exponents p_2 and p_3 are defined are rotated from (x^2, x^3) to $(X_{(z)}^2, X_{(z)}^3)$ at each fixed point z.

(c) $0 \leq \alpha \leq 1/2$. Then, the metric takes the form

$$ds^2 = \varepsilon_0^3 d\tau^2 - \varepsilon_1^3 \tau^{2p_1} dz^2 - \varepsilon_2^3 \tau^{2p_2} (dX_{(z)}^2)^2 - \varepsilon^3 3\tau^{2p_3} (dX_{(z)}^3)^2, \qquad (4.73)$$

as $t \to 0^+$, but now τ and p_k are defined by

$$\tau \equiv t^{\alpha^2 - \alpha + 1}, \quad p_1 \equiv \frac{\alpha(\alpha - 1)}{\alpha^2 - \alpha + 1},$$

$$p_2 \equiv \frac{\alpha}{\alpha^2 - \alpha + 1}, \quad p_3 \equiv \frac{1 - \alpha}{\alpha^2 - \alpha + 1}. \qquad (4.74)$$

The comparison of the above limit with the one in the collinear case shows that the nature of the singularities on the hypersurface $t = 0$ is changed in the present case. In the non-collinear case it is anastigmatic, while in the collinear case it is astigmatic. Moreover, the Kasner axes in the present case are also rotated. When $\alpha = 0$, Eq. (4.74) gives $p_1 = 0$. That is, the solutions with $\alpha = 0$ are free of curvature singularities on the hypersurface $t = 0$.

(d) $\alpha > 1/2$. Then, we find

$$ds^2 = \varepsilon_0^4 d\tau^2 - \varepsilon_1^4 \tau^{2p_1} dz^2 - \varepsilon_2^4 \tau^{2p_2} \left(dx^2\right)^2 - \varepsilon_3^4 \tau^{2p_3} \left(dx^3\right)^2, \qquad (4.75)$$

where τ and p_k are given by Eqs. (4.74) and (4.75). Equation (4.75) shows that in the present case the Kasner axes remain the same as these in the collinear case, but the Kasner exponents are changed. When $1/2 < \alpha < 1$, we have $p_1 < 0$. Thus, the nature of the singularities is anastigmatic (note that in the collinear case, it is astigmatic). When $\alpha > 1$, we have $p_1 > 0$, which means that the nature of the singularities is astigmatic, as it is in the collinear case. When $\alpha = 1$, we have $p_1 = 0$. Thus, the solutions with $\alpha = 1$ are free of spacetime curvature singularities on the hypersurface $t = 0$, too. The above analysis shows that the non-diagonal solutions given by Eqs. (4.56)–(4.58) are free of spacetime curvature singularities on the hypersurface t = 0, if any one of the following conditions holds:

$$\text{(i) } a + \delta_1 = -1, \quad \text{(ii) } a + \delta_1 = 0, \quad \text{or} \quad \text{(iii) } a + \delta_1 = 1. \qquad (4.76)$$

Comparing Eq. (4.71) with Eq. (4.76), we find that the solutions with $\alpha = 0$ or -1 are free of curvature singularities in both of the collinear and non-collinear cases, whereas the solutions with $\alpha = 1$ are free of spacetime curvature singularities only in the non-collinear case. The above difference should be obviously attributed to the presence of the ×-polarization mode of the plane gravitational wave moving toward the right-hand side. In other words, the interaction between different polarization modes can change the nature of the singularities on the hypersurface $t = 0$ and turn some space-like singularities into Cauchy horizons.

4.2.3. Specific solutions

To illustrate further the features of the solutions presented in the last subsection, we consider some special cases corresponding to different values of the free parameters n, m and a.

Case A: $\delta_1 = \delta_2 = 1/2$: In this case, from Eq. (4.61) we find

$$n = m = 1. \qquad (4.77)$$

The previous analysis shows that in this case the solutions in general represent the collision of plane gravitational shock + impulsive waves. Setting

$\delta_1 = \delta_2 = 1/2$ in Eqs. (4.56)–(4.58) and considering Eq. (3.42), we find that the solutions are now given by

$$\chi = [(1 - \eta^2)(1 - \mu^2)]^a \frac{A}{B},$$

$$q_2 = -2q \frac{[(1 - \eta^2)(1 - \mu^2)]^{2a}}{B}(\eta - \mu), \tag{4.78}$$

$$e^{-M} = \frac{[(1 - \eta^2)(1 - \mu^2)]^{a^2 - 1/4} A}{XY(1 + \eta)^{2a}(1 + \mu)^{2a}},$$

where the functions A and B now are defined by

$$A \equiv (1 - \eta^2)(1 - \eta)^{2a}(1 + \mu)^{2a} + q^2(1 - \mu^2)(1 + \eta)^{2a}(1 - \mu)^{2a},$$
$$B \equiv (1 + \eta)^2(1 - \eta)^{2a}(1 + \mu)^{2a} + q^2(1 + \mu)^2(1 + \eta)^{2a}(1 - \mu)^{2a}. \tag{4.79}$$

It can be shown that this subclass of solutions belongs to the Ernst–Garcia–Hauser solutions (Ernst, Garcia and Hauser, 1987a, 1987b, 1988). To study these solutions as a whole is very complicated. In the following, we consider only some representative cases.

Case A.1: $a = 0$. In this case, we find that the non-vanishing Weyl scalars are given by

$$\Psi_0(u, v) = \Psi_0^{IV}(uH(u), v)H(v) + \frac{1}{(1 + q^2)Y}[(1 - q^2) - i2q]\delta(v),$$

$$\Psi_4(u, v) = \Psi_4^{IV}(u, vH(v))H(u) + \frac{1}{(1 + q^2)DX}[(1 + q^4)(1 - v^2H(v))$$
$$\qquad + 2q^2(1 + 3v^2H(v)) + 14q(1 - q^2)v^2H(v)]\delta(u), \tag{4.80}$$

$$\Psi_2(u, v) = \Psi_2^{IV}(u, v)H(u)H(v),$$

where

$$\Psi_0^{II}(v) = \Psi_0^{IV}(0, v) = 0, \quad \Psi_4^{III}(u) = \Psi_4^{IV}(u, 0) = 0,$$
$$D(v) \equiv [(1 + q^2)^2(1 - v^2H(v))^2 + 16q^2v^2H(v)]^{1/2}. \tag{4.81}$$

Equations (4.80) and (4.81) show that this model represents the collision of two impulsive plane gravitational waves. Thus, the corresponding solution is either the Khan–Penrose solution (Khan and Penrose, 1971) or the Nutku–Halil solution (Nutku and Halil, 1977). A detained analysis of the polarizations of these two plane gravitational impulsive waves shows that it is the Nutku–Halil solution, since the polarization angles for these two incoming impulsive waves are different, although they are constants.

This solution was studied in detail by Chandrasekhar and Ferrari but with a different form (Chandrasekhar and Ferrari, 1984). It was found that the solution is singular on the hypersurface $t = 0$.

Case A.2: $a = 1/2$. In this case, the non-vanishing Weyl scalars are given by

$$\Psi_0(u, v) = \Psi_0^{IV}(uH(u), v)H(v) + \frac{1}{Y(1+q^2)}[(1-q^2) - i2q]\delta(v),$$

$$\begin{aligned}\Psi_4(u, v) = {}& \Psi_4^{IV}(u, vH(v))H(u) + \frac{1}{CDX}\{[(1-vH(v))^4 \\ & + 2q^2[1 + 2v^2H(v) - 3v^4H(v)] + q^4(1+vH(v))^4] \qquad (4.82) \\ & + i4qv^2H(v)[(1-v)^2 - q^2(1+v)^2]\}\delta(u),\end{aligned}$$

$$\Psi_2(u, v) = \Psi_2^{IV}(u, v)H(u)H(v),$$

where

$$\begin{aligned}\Psi_0^{II}(v) = {}& \Psi_0^{IV}(0, v) \\ = {}& -\frac{3(1-v^2)}{(1+q^2)^{1/2}C^2F}\{[16q^4 - (1-q^2)^4 + (9-14q^2)v] \\ & - i2q(1+q^2)[(1+q^2)(5+v^2)v - 2(1-q^2)(1+2v^2)] \\ & - [(1+q^2)(1+6v^2+v^4) - 4(1-q^2)(1+v^2)v]\}, \\ & \qquad\qquad\qquad\qquad\qquad\qquad\qquad\qquad\qquad\qquad (4.83) \\ \Psi_4^{III}(u) = {}& \Psi_4^N(u0) = \frac{3(1+u)}{(1-u^2)^2}, \\ C \equiv {}& (1-vH(v))^2 + q^2(1+vH(v))^2, \\ D \equiv {}& [C^2 + 16q^2v^2H(v)]^{1/2}, \\ F \equiv {}& [(1+q^2)(1+6^2+v^4) - 4(1-q^2)(1+v^2)v]^{3/2},\end{aligned}$$

and

$$\begin{aligned}\Psi_2^{IV}(u, v) = {}& \frac{(1+\eta)^2(1+\mu)^2}{XYA^2}\{(1-\eta)^3 - q^2(1-\mu)[2 - (1-\eta)(\eta+\mu) \\ & + 2\eta^2\mu - q^2(1-3\mu^2+2\mu^3)] \\ & + \frac{iq(1+\eta)(1+\mu)}{B[(1-\eta)^3(1+\mu) + q^2(1-\mu)^3(1+\eta)]}\end{aligned}$$

$$\times [(1 - \eta)^6 (1 + \eta)(1 + \mu)(2 + \eta - 3\mu)$$
$$+ q^4 (1 - \eta)(1 - \mu)^3 (6 - 5\eta + 2\eta^2 - 7\eta^3 - \mu - 16\eta\mu + 29\eta^2\mu$$
$$+ 2\mu^2 - 15\eta\mu^2 + 6\eta^2\mu^2 - 5\eta^3\mu^2 + 5\eta^3 - \eta^2\mu^3)$$
$$- q^6 (1 + \eta)(1 + \mu)(1 - \mu)^6 (2 - 3\eta + \mu)]\}. \tag{4.84}$$

Equations (4.82) and (4.84) show that in this case the solution represents the collision of two impulsive + shock waves, one of which is constantly polarized and the other is variably polarized.

From Eqs. (4.71) and (4.76), we find that this solution is free of space-time singularities on the hypersurface $t = 0$ only when $q \neq 0$, and when $q = 0$ it is always singular. This is shown clearly by Eq. (4.84), from which we can see that when $q \neq 0$ the Weyl scalar Ψ_2^{IV} remains finite as $\eta \to 1$, but when $q = 0$, Ψ_2^{IV} becomes unbounded. Actually, setting $q = 0$, Eq. (4.84) becomes

$$\Psi_2^{\mathrm{IV}}(q = 0) = \frac{1}{XY(1 - \eta)}. \tag{4.85}$$

Case B: $\delta_1 = \delta_2 = \sqrt{6}/4$. In this case, from Eq. (4.61) we find

$$n = 2, \quad m = 1. \tag{4.86}$$

Then, the corresponding solutions are given by

$$\chi = [(1 - \eta^2)(1 - \mu^2)]^{a + \frac{1}{2}} \left[\frac{1 - \eta}{1 + \eta}\right]^k \left[\frac{1 - \mu}{1 + \mu}\right]^k \frac{A}{B},$$

$$q_2 = -\frac{2q}{B}(\eta - \mu)[(1 - \eta^2)(1 - \mu^2)]^{2a}[(1 - \eta)(1 - \mu)]^{2k}, \tag{4.87}$$

$$e^{-M} = \frac{[(1 - \eta^2)(1 - \mu^2)]^\rho A}{X^{3/2} Y [(1 + \eta)(1 + \mu)]^{4ak}},$$

but now the functions A and B are defined as

$$A \equiv (1 - \eta^2)^{2k}[(1 - \eta)(1 + \mu)]^{2a}$$
$$+ q^2 (1 - \mu^2)^{2k}[(1 + \eta)(1 - \mu)]^{2a},$$
$$B \equiv (1 - \eta^2)^{2k}[(1 - \eta)(1 + \mu)]^{2a}(1 + \eta)(1 - \mu)$$
$$+ q^2 (1 - \mu^2)^{2k}[(1 + \eta)(1 - \mu)]^{2a}(1 - \eta)(1 + \mu), \tag{4.88}$$

where

$$k = \frac{\sqrt{6}}{4}, \quad \rho \equiv (a + k)(a + k - 1). \tag{4.89}$$

This class of solutions in general represents the collision of a variably polarized shock + impulsive gravitational wave with a constantly polarized shock wave.

Case B.1: $a = 0$. In this case, we have

$$\Psi_0(u,v) = \Psi_0^{IV}(uH(u),v)H(v) + \frac{1}{Y(1+q^2)}\left[(1-q^2) - i2q\right]\delta(v),$$

$$\Psi_4(u,v) = \Psi_4^{IV}(u,vH(v))H(u), \tag{4.90}$$

$$\Psi_2(u,v) = \Psi_2^{IV}(u,v)H(u)H(v),$$

and

$$\Psi_0^{II}(v) = \Psi_0^{IV}(0,v) = 0, \quad \Psi_4^{III}(u) = \Psi_4^{IV}(u,0) = 4kY^{-4}. \tag{4.91}$$

Equations (4.90) and (4.91) show that in this case the solution represents the collision of a constantly polarized shock wave and an impulsive wave. Comparing it with Case A.1, we find that one of the two incoming impulsive gravitational waves in the Nutku–Halil solution now is replaced by a constantly polarized shock wave.

To study the singularity behavior of the solution on the hypersurface $t = 0$, we find that the Weyl scalar Ψ_2^{IV} is given by

$$\Psi_2^{IV}(u,v) = \frac{4ku}{XYt^2A^2B}\{B[(1-\eta^2)^{4k}[2 - (1-k)\eta^2 - (1+k)\mu^2]$$

$$- 6kq^2(\eta^2 - \mu^2)t^{4k}$$

$$- q^4(1-\mu^2)^{4k}[2 - (1-k)\mu^2 - (1+k)\eta^2]]$$

$$- iqt^{2k}(1-\eta^2)^{4k}[k^{-1}(\eta^2 - \mu^2)(1+\eta)(1-\mu)$$

$$+ (2 + 3\eta + 6\eta^2 + 3\eta^3) - \mu(3 + 7\eta + \eta^2 - \eta^3)$$

$$- \mu^2(1 + 2\eta + 5\eta^2) + 2\mu^3(1+\eta)]\}. \tag{4.92}$$

Obviously, as $t \to 0^+$, the Weyl scalar Ψ_2^{IV} becomes unbounded. That is, similar to the Nutku–Halil solution (Chandrasekhar and Ferrari, 1984), a spacetime singularity is finally developed on the hypersurface $t = 0$ due to the mutual focus of the two colliding gravitational waves.

Case B.2: $a + \delta_1 = 1$. In this case, the non-vanishing Weyl scalars are given by

$$\Psi_0(u,v) = \Psi_0^{IV}(uH(u),v)H(v) + \frac{1}{Y(1+q^2)}[(1-q^2) - i2q]\delta(v),$$

$$\Psi_4(u,v) = \Psi_4^{IV}(u,vH(v))H(u), \tag{4.93}$$

$$\Psi_2(u,v) = \Psi_2^{IV}(u,v)H(u)H(v),$$

and

$$\Psi_0^{II}(v) = \Psi_0^{IV}(0, v) = \frac{1}{2(1 - v^2) A^2 D} \{[O(v) - q^2 P(v) - q^4 P(-v)$$

$$+ q^6 O(-v)] - i2q(1 - v^2)^{1+2k}[Q(v) - 2q^2 Z(v) - q^4 Q(-v)]\},$$

$$\Psi_4^{III}(u) = \Psi_4^{IV}(u, 0) = 4Y^{-4}[k + 9au^3 + 24a^2ku^5 + 2a(4a^2 - 1)u^6],$$

$$(4.94)$$

where

$$O(x) \equiv (1 + x)^{12k}(1 - x)^{12} \Big\{ 3(4 + 7x + 2x^2)$$

$$- 2k[6 + 24x + 19x^2 - 16kx(1 + 2x - kx)]\},$$

$$P(x) \equiv (1 + x)^{8k+3}(1 - x)^{4k+7} \{3(20 + 121x + 22x^2 + 7x^3 + 6x^4)$$

$$- 2k[(30 + 360x + 113x^2 + 24x^3 + 57x^4)$$

$$- 16kx(11 + 10x + x^2 + 6x^3) + 16k^2x^2(5 + 3x^2)]\}, \qquad (4.95)$$

$$Q(x) \equiv (1 + x)^{8k}(1 - x)^8 \{3(8 + 23x + 4x^2 - 9x^3)$$

$$- 4k[(6 + 36x + 19x^2 - 12x^3) - 4kx(5 + 8x - x^2) + 16k^2x^2]\},$$

$$Z(x) \equiv x(1 - x^2)^{4(k+1)}[3(41 + 9x^2) - 16k(15 + 3x^2) + 16k^2(7 + x^2)],$$

$$D(x) \equiv \left[A^2 + 16q^2x^2 (1 - x^2)^{4(1+k)}\right]^{1/2}.$$

Thus, unlike the last case, the Nutku–Halil impulsive wave incident in Region II now is replaced by a variably polarized shock + impulsive wave. The solution in the present case is free of spacetime singularity on the hypersurface $t = 0$ only when $q \neq 0$, and otherwise it is always singular. This can be seen from the Weyl scalar Ψ_2^{IV}, which is given by

$$\Psi_2^{IV}(u, v) = \frac{4u}{XYBA^2(1 + \eta)(1 + \mu)} \{B[J(\eta, \mu) - 6q^2 K(\eta, \mu) - q^4 J(\mu, \eta)]$$

$$- i2q(1 + \eta)^{2k+1}(1 + \mu)^{2k+1}[L(\eta, \mu) + 2q^2 N(\eta, \mu)$$

$$+ q^4 L(\mu, \eta)]\}, \qquad (4.96)$$

where

$$J(x, y) \equiv (1 + y)^4(1 - x)^3(1 + x)^{8k}[(1 + x)(x + y)$$

$$+ k(1 - x)(2 + 3x - y) - 2k^2(1 - x)(x - y)],$$

$$K(x,y) \equiv (x-y)(1-x^2)(1-y^2)(1+x)^{4k+1}(1+y)^{4k+1}$$
$$\times [(x+y) - 2k(1-k)(1-x)(x-y)],$$
$$L(x,y) \equiv (1+y)^3(1-x)^3(1+x)^{8k}\{(1+2x-2y-xy)(x+y)$$
$$+ k(1-x)(1-y)[(2+5x-3y) - 4k(x+y)]\}, \qquad (4.97)$$
$$N(x,y) \equiv (1-x^2)(1-y^2)(1+x)^{4k}(1+y)^{4k}\{(x+y)[(1+xy)^2$$
$$- 2(x^2+y^2)] + k(1-x)(x-y)[(1+x)(1+y)(2-x-y)$$
$$- 3(x-y)^2 + 4k(x-y)^2]\}.$$

When $q = 0$, Eq. (4.96) reduces to

$$\Psi_2^{\text{IV}}(u,v) = \frac{4u(1-\mu)}{XYt^2}[(1+\eta)(\eta+\mu) + k(1-\eta)(2+3\eta-\mu)$$
$$- 2k^2(1-\eta)(\eta-\mu)], \qquad (4.98)$$

which becomes unbounded as $t \to 0^+$.

4.2.4. Polarizations of colliding plane gravitational waves

To study the polarizations of colliding plane gravitational waves, in this subsection we consider only two special cases of the solutions presented in the last subsection.

Case 1: $n = m = 1$, $a = 0$. This is the case discussed in Subsection 4.2.3. The non-vanishing Weyl scalars are given by Eq. (4.80). For the sake of simplicity, we consider only the impulsive part. Thus, according to Eqs. (2.38) and (2.46) we have

$$\varphi_0^{\text{Im}} = \frac{1}{2}\tan^{-1}\left(\frac{2q}{1-q^2}\right),$$
$$\varphi_4^{\text{Im}} = \frac{1}{2}\tan^{-1}\left(\frac{4q(1-q^2)v^2}{(1+q^4)(1-v^2) + 2q^2(1+3v^2)}\right)H(v). \qquad (4.99)$$

On the other hand, from Eq. (4.78) we find that on the hypersurface $v = 0$ we have $q_2 = 0$ (or equivalently, $W = 0$). Consequently, Eq. (3.61) yields,

$$\varphi_0^{\text{Im}(0)} = 0. \qquad (4.100)$$

Thus, we have

$$\theta_0^{\text{Im}} = \varphi_0^{\text{Im}} - \varphi_0^{\text{Im}(0)} = \frac{1}{2}\tan^{-1}\left(\frac{2q}{1-q^2}\right), \qquad (4.101)$$

which means that the polarization of the impulsive plane gravitational wave with support on the hypersurface $v = 0$ does not change even after the collision.

Substituting Eqs. (4.43) and (4.78) into Eq. (3.59), and then integrating it, for $v > 0$ we find,

$$\varphi_4^{\text{Im}(0)} = \frac{1}{2}\left(\tan^{-1}\frac{\left(1+q^2\right)v - \left(1-q^2\right)}{2q}\right.$$
$$\left. -\tan^{-1}\frac{\left(1+q^2\right)v + \left(1-q^2\right)}{2q} + C_0\right). \tag{4.102}$$

From the condition that when $v \to 0$, $\varphi_4^{\text{Im}(0)} = 0$, we have

$$\tan C_0 = \frac{4q\left(1-q^2\right)}{8-\left(3-q^2\right)^2}. \tag{4.103}$$

Thus, Eqs. (4.102) and (4.103) yield,

$$\varphi_4^{\text{Im}(0)} = \frac{1}{2}\tan^{-1}\left(\frac{4q\left(1-q^2\right)v^2}{\left(1+q^4\right)\left(1-v^2\right)+2q^2\left(1+3v^2\right)}\right)H(v). \tag{4.104}$$

Equations (3.60), (4.99) and (4.104), on the other hand, give

$$\theta_4^{\text{Im}} = 0. \tag{4.105}$$

This means that the impulsive wave with support on the hypersurface $u = 0$ has zero polarization angle relative to the parallelly transported basis even after the collision. Thus, the solution given by Eqs. (4.78) and (4.79) with $a = 0$ represents the collision of two impulsive plane gravitational waves with different polarization angles. This must be the Nutku–Halil solution. Actually, if we make a rotation in the (x^2, x^3)-plane with the angle given by

$$\varphi \equiv \frac{1}{2}\left(\theta_0^{\text{Im}} - \theta_4^{\text{Im}}\right) = \frac{1}{4}\tan^{-1}\left(\frac{2q}{1-q^2}\right), \tag{4.106}$$

we find that the corresponding solution takes the exact form of the Nutku–Halil solution used by Chandrasekhar and Ferrari (1984).

Case 2: $n = m = 1$, $a = 1/2$. This case was also considered in the last subsection. The non-vanishing Weyl scalars are given by Eq. (4.82). Comparing Eq. (4.80) with Eq. (4.82) we find that the impulsive part of Ψ_0 in the above two cases are the same. From Eq. (4.78) we find that the impulsive part of Ψ_0 has the same polarization angle θ_4^{Im} as the one given in the last case.

Integrating Eq. (3.59) along the hypersurface $u = 0$, on the other hand, we find

$$\varphi_4^{\mathrm{Im}(0)} = \frac{1}{2}\tan^{-1}\left(\frac{4qv^2\left[(1-v)^2 - q^2(1+v)^2\right]}{(1-v)^4 + 2q^2\left(1 + 2v^2 - 3v^4\right) + q^4(1+v)^4}\right)H(v).$$

(4.107)

Then, Eqs. (2.38), (4.59), (4.82) and (4.107) give

$$\theta_4^{\mathrm{Im}} = 0.$$

(4.108)

Thus, similar to the last case, the polarization angle for the impulsive part of Ψ_4 is zero. That is, the impulsive part of the Ψ_4 wave does not change its polarization direction when it collides and interacts with the oppositely moving plane gravitational wave. The above results are consistent with our general conclusions obtained in Section 3.4, from there we can see that the reason that the impulsive part of colliding plane gravitational wave does not change its polarization direction is that in the present case this part does not interact with any of the others [sec Eq. (3.69)].

4.3. Collisions of Two Non-collinearly Polarized Gravitational Plane Waves

In the last section, by adding one soliton into the diagonal solutions discussed in Section 4.1, a four-parameter class of solutions was obtained, which in general represents the collision of two plane gravitational waves, one of which is constantly polarized, while the other is variably polarized. A natural generalization of the above solutions is to add one more soliton into the seed solutions.

4.3.1. Two-soliton solutions

By choosing the trajectories of the two poles as

$$\mu_1 = 1 - z - [(1-z)^2 - t^2]^{1/2} = (1-\eta)(1+\mu),$$
$$\mu_2 = -1 - z - [(-1-z)^2 - t^2]^{1/2} = -(1-\eta)(1-\mu),$$

(4.109)

we find that the two-soliton solutions with the seed given by Eqs. (4.23) and (4.24) are given by (Wang, 1991a)

$$\chi = \frac{A}{B}\chi^{(0)}, \quad q_2 = \frac{C}{B}(\chi^{(0)})^2,$$
$$f = C_0\frac{(1-\eta)^{\rho_1}(1-\mu)^{\rho_2}(1+\eta)^{r_1}(1+\mu)^{r_2}}{(\eta+\mu)^{(\delta_+ +1)^2}(\eta-\mu)^{(\delta_- +1)^2}}A,$$

(4.110)

where C_0 is a constant,

$$\chi^{(0)} = [(1-\eta^2)(1-\mu^2)]^a \left(\frac{1-\eta}{1+\eta}\right)^{\delta_1} \left(\frac{1-\mu}{1+\mu}\right)^{\delta_2}, \tag{4.111}$$

and

$$
\begin{aligned}
A \equiv {}& (1-\eta^2)(1-\mu)^{2(a+\delta_2)}(1+\mu)^{2(a-\delta_2)}[(1-\eta)^{2(a+\delta_1)} \\
& + Q_1 Q_2 (\eta-\mu)^{2\delta} - (\eta+\mu)^{2\delta} + (1+\eta)^{2(a-\delta_1)}]^2 \\
& + (1-\mu^2)(1-\eta)^{2(a+\delta_1)}(1+\eta)^{2(a-\delta_1)}[Q_1(\eta+\mu)^{2\delta} + (1+\mu)^{2(a-\delta_2)} \\
& - Q_2(\eta-\mu)^{2\delta} - (1-\mu)^{2(a+\delta_2)}]^2, \\
B \equiv {}& (1-\mu)^{2(a+\delta_2)}(1+\mu)^{2(a-\delta_2)}[(1+\eta)(1-\eta)^{2(\alpha+\delta_1)} \\
& - Q_1 Q_2 (1-\eta)(\eta-\mu)^{2\delta} - (\eta+\mu)^{2\delta_+}(1+\eta)^{2(a-\delta_1)}]^2 \\
& + (1-\eta)^{2(a+\delta_1)}(1+\eta)^{2(a-\delta_1)}[Q_1(1-\mu)(\eta+\mu)^{2\delta} + (1+\mu)^{2(a-\delta_2)} \\
& + Q_2(1+\mu)(\eta-\mu)^{2\delta} - (1-\mu)^{2(a+\delta_2)}]^2, \\
C \equiv {}& -2\{Q_1(\eta+\mu)^{2\delta_++1}[(1+\mu)^{2(a-\delta_2)}(1-\eta)^{2(a+\delta_1)} \\
& + Q_2^{\,2}(\eta-\mu)^{4\delta_-}(1-\mu)^{2(a+\delta_2)}(1+\eta)^{2(a-\delta_1)}] \\
& + Q_2(\eta-\mu)^{2\delta_-+1}[(1-\mu)^{2(a+\delta_2)}(1-\eta)^{2(a+\delta_1)} \\
& + Q_1^2(\eta+\mu)^{4\delta_+}(1+\mu)^{2(\alpha-\delta_2)}(1+\eta)^{2(a-\delta_1)}]\},
\end{aligned}
\tag{4.112}
$$

where Q_1 and Q_2 are arbitrary constants, and

$$\rho_A \equiv (a+\delta_A)(a+\delta_A-2) - \frac{1}{4}, \quad r_A \equiv (a-\delta_A)(a-\delta_A-2) - \frac{1}{4},$$
$$(A = 1, 2). \tag{4.113}$$

Note that when $\eta \to \pm\mu$ the functions A, B and C defined by Eq. (4.112) become unbounded unless $\delta_\pm \geq 0$. Therefore, for the extension of these solutions beyond the hypersurfaces $\eta = \pm\mu$, we restrict the solutions given in the form of Eqs. (4.110)–(4.111) only to the cases where $\delta_\pm \geq 0$. Otherwise, we write these solutions in the following different forms.

(i) $\delta_+ > 0$, $\delta_- < 0$: In this case, the solutions given by Eqs. (4.110)–(4.111) are written in the form

$$\chi = \frac{A_{(1)}}{B_{(1)}}\chi^{(0)}, \quad q_2 = \frac{C_{(1)}}{B_{(1)}}(\chi^{(0)})^2,$$

$$f = C_0 \frac{(1-\eta)^{\rho_1}(1-\mu)^{\rho_2}(1+\eta)^{T_1}(1+\mu)^{1_2}}{(\eta+\mu)^{(\delta,+1)^2}(\eta-\mu)^{(\delta-1)^2}} A_{(1)}, \tag{4.114}$$

where

$$A_{(1)} \equiv (\eta - \mu)^{-4\delta_-} A, \quad B_{(1)} \equiv (\eta - \mu)^{-4\delta_-} B, \quad C_{(1)} \equiv (\eta - \mu)^{-4\delta_-} C, \tag{4.115}$$

and the functions A, B and C are given by Eq. (4.112).

(ii) $\delta_+ < 0$, $\delta_- > 0$: Then, in this case we write these solutions in the form,

$$\chi = \frac{A_{(2)}}{B_{(2)}} \chi^{(0)}, \quad q_2 = \frac{C_{(2)}}{B_{(2)}} (\chi^{(0)})^2,$$

$$f = C_0 \frac{(1-\eta)^{\rho_1} (1-\mu)^{\rho_2} (1+\eta)^{r_1} (1+\mu)^{r_2}}{(\eta + \mu)^{(\delta_+ - 1)^2} (\eta - \mu)^{(\delta_- + 1)^2}} A_{(2)}, \tag{4.116}$$

where the functions $A_{(2)}, B_{(2)}$ and $C_{(2)}$ are related to the functions A, B and C given by Eq. (4.112) via the relations,

$$A_{(2)} = (\eta + \mu)^{-4\delta_+} A, \quad B_{(2)} = (\eta + \mu)^{-4\delta_+} B, \quad C_{(2)} = (\eta + \mu)^{-4\delta_+} C. \tag{4.117}$$

(iii) $\delta_\pm < 0$: In this case, we have

$$\chi = \frac{A_{(3)}}{B_{(3)}} \chi^{(0)}, \quad q_2 = \frac{C_{(3)}}{B_{(3)}} (\chi^{(0)})^2,$$

$$f = C_0 \frac{(1-\eta)^{\rho_1} (1-\mu)^{\rho_2} (1+\eta)^{r_1} (1+\mu)^{r_2}}{(\eta + \mu)^{(\delta_+ - 1)^2} (\eta - \mu)^{(\delta_- - 1)^2}} A_{(3)}, \tag{4.118}$$

where

$$A_{(3)} \equiv (\eta + \mu)^{-4\delta_+} (\eta - \mu)^{-4\delta_-} A, \quad B_{(3)} \equiv (\eta + \mu)^{-4\delta_+} (\eta - \mu)^{-4\delta_-} B,$$
$$C_{(3)} \equiv (\eta + \mu)^{-4\delta_+} (\eta - \mu)^{-4\delta_-} C. \tag{4.119}$$

Equations (4.110)–(4.119) show that in all the above cases, the function f takes the form

$$f = \frac{\tilde{f}(\eta, \mu)}{(\eta + \mu)^{\alpha_+} (\eta - \mu)^{\alpha_-}}, \tag{4.120}$$

for example, in Eq. (4.110) we have $\alpha_+ = (\delta_+ + 1)^2$ and $\alpha_- = (\delta_- + 1)^2$, etc.

The analysis carried out in the last two sections shows that for the function f having the form of Eq. (4.120) the parameters n and m must be chosen so that

$$n = (2 - \alpha_+)^{-1}, \quad m = (2 - \alpha_-)^{-1}. \tag{4.121}$$

Without loss of generality, in the following we consider the above solutions only for the cases in which $\delta_\pm \geq 0$. The above solutions in the other cases will have similar physical interpretations and properties. For more details, we refer the readers to Wang (1991a).

When $\delta_{\pm} \geq 0$, Eq. (4.121) becomes

$$n = [2 - (\delta_+ + 1)^2]^{-1}, \quad m = [2 - (\delta_- + 1)^2]^{-1}. \tag{4.122}$$

As usual, we take the solutions given by Eqs. (4.110)–(4.111) as valid only in Region IV, then extend them back to the pre-collision regions by means of the Khan–Penrose substitutions (3.11). Such extended solutions represent the collisions of purely gravitational waves only in the cases where $n, m \geq 1$, which in the present case implies that

$$0 \leq \delta_{\pm} < \sqrt{2} - 1. \tag{4.123}$$

From the discussion given in the last two sections, we can see that the extended two-soliton solutions represent a variety of models of the collision of two pure gravitational plane waves. For example, when $n = 1$ and $m = 2$ the solutions represent the collision of a variably polarized shock and a variably polarized shock + impulsive wave, and so on.

Comparing with the seed solutions, we can see that in the two-soliton case both of the two constantly (collinearly) polarized gravitational plane waves of the seed solutions are generalized to variably (non-collinearly) polarized ones, whereas in the one-soliton case only one of them is generalized to a variably polarized wave, while the other is still constantly polarized.

4.3.2. Formation and nature of spacetime singularities

To study the behavior of the spacetime singularities of the above solutions near the focusing hypersurface $t = 0$, we start with their collinear limit, as we did in the one-soliton case. Setting $Q_1 = Q_2 = 0$, we find that the solutions given by Eqs. (4.110)–(4.111) reduce to

$$\chi = \left(\frac{1-\eta}{1+\eta}\right) \chi^{(0)} = \left((1-\eta^2)(1-\mu^2)\right)^a \left(\frac{1-\eta}{1+\eta}\right)^{\delta_1+1} \left(\frac{1-\mu}{1+\mu}\right)^{\delta_2},$$

$$q_2 = 0, \tag{4.124}$$

$$f = C_0 \frac{(1-\eta)^{b_1'}(1-\mu)^{b_2'}(1+\eta)^{c_i'}(1+\mu)^{c_2'}}{(\eta+\mu)^{(\delta_+ +1)^2}(\eta-\mu)^{(\delta_- +1)^2}},$$

where

$$b_1' = (a + \delta_1 + 1)^2 - \frac{1}{4}, \quad b_2' = (a + \delta_2)^2 - \frac{1}{4},$$

$$c_1' = (a - \delta_1 - 1)^2 - \frac{1}{4}, \quad c_2' = (a - \delta_2)^2 - \frac{1}{4}. \tag{4.125}$$

But, these are the solutions given by Eqs. (4.23) and (4.24), if we replace δ_1 by $\delta_1 + 1$. Thus, replacing α by $\alpha' = a + \delta_1 + 1$ in Eqs. (4.50)–(4.52), we obtain the asymptotic behavior of the solutions given by Eqs. (4.124) and (4.125). Specifically, the conditions of Eq. (4.53) now become

$$(a + \delta_1 + 1)^2 = \frac{1}{4}, \tag{4.126}$$

which is equivalent to

$$\alpha = -\frac{1}{2}, \quad \text{or} \quad \alpha = -\frac{3}{2}. \tag{4.127}$$

On the other hand, in the non-collinear case we have $Q_1 Q_2 \neq 0$, which allows us to distinguish the following cases.

(α) $\alpha < -1$. Then, the solutions given by Eqs. (4.110)–(4.111) have the same limit as their diagonal solutions given by (4.124) as $t \to 0^+$. The solutions with $\alpha = -3/2$ are especially free of spacetime singularities on the hypersurface $t = 0$ in both collinear and non-collinear cases. When $\alpha < -3/2$, we have $p_1 > 0$. That is, the corresponding singularities are astigmatic. When $-3/2 < \alpha < -1$, we have $p_1 < 0$, and the corresponding singularities are anastigmatic.

(β) $-1 \leq \alpha \leq -1/2$. In this case, the solutions have the limit

$$ds^2 \simeq \varepsilon_0^{(2)} d\tau^2 - \varepsilon_1^{(2)} \tau^{2p_1} dz^2 - \varepsilon_2^{(2)} \tau^{2p_2} (dX_{(z)}^2)^2$$
$$- \varepsilon_3^{(2)} \tau^{2p_3} (dX_{(z)}^3)^2, \tag{4.128}$$

as $t \to 0^+$, where τ and p_i are the same as they are defined in the corresponding diagonal case. Equation (4.128) shows that the Kasner exponents p_i and the time τ are not changed relative to the corresponding diagonal case, but the Kasner axes along which the p_2 and p_3 are defined are rotated from (x^2, x^3) to $(X_{(z)}^2, X_{(z)}^3)$. Since in the present case $p_1 < 0$, the nature of the singularities on the hypersurface $t = 0$ is anastigmatic, as it is in the collinear case. When $\alpha = -1/2$, we have $p_1 = 0$. Hence, the solutions with $\alpha = -1/2$ are free of spacetime curvature singularities on $t = 0$.

(γ) $-1/2 < \alpha < 1/2$. In this case, the solutions have the limit

$$ds^2 = \varepsilon_0^{(3)} d\tau^2 - \varepsilon_1^{(3)} \tau^{2p_1} dz^2 - \varepsilon^{(3)} 2\tau^{2p_2} (dX_{(z)}^2)^2$$
$$- \varepsilon_3^{(3)} \tau^{2p_3} (dX_{(z)}^3)^2, \tag{4.129}$$

as $t \to 0^+$, with τ and p_i given exactly by Eqs. (4.50) and (4.51). Recall that in the collinear case these solutions have the limit of Eqs. (4.49) and (4.51). But, now the same limit is obtained with the replacement of α by $\alpha' = \alpha + \delta_1 + 1$. Thus, we find that in the present case the interaction between different polarization modes of the two colliding plane gravitational waves changes both the Kasner exponents and axes. The nature of singularities in the present case is anastigmatic, whereas it is astigmatic in the collinear case.

(δ) $1/2 \leq \alpha \leq 1$. Then, the solutions have the limit

$$ds^2 \simeq \varepsilon_0^{(4)} d\tau'^2 - \varepsilon_1^{(4)} \tau'^{2p'_1} dz^2 - \varepsilon_2^{(4)} \tau^{2p'_2} (dX_{(z)}^2)^2$$
$$- \varepsilon_3^{(4)} \tau'^{2p'_3} (dX_{(z)}^3)^2, \tag{4.130}$$

as $t \to 0^+$, where

$$\tau' \equiv t^{(\alpha-1)^2+3/4}, \tag{4.131}$$

and

$$p'_1 \equiv \frac{(\alpha-1)^2 - \frac{1}{4}}{(\alpha-1)^2 + \frac{3}{4}}, \quad p'_2 \equiv \frac{\alpha - \frac{1}{2}}{(\alpha-1)^2 + \frac{3}{4}}, \quad p'_3 \equiv \frac{\frac{3}{2} - \alpha}{(\alpha-1)^2 + \frac{3}{4}}. \tag{4.132}$$

It is easy to show that the exponents p_i defined by Eq. (4.132) also satisfy the Kasner relations (3.92).

As in the last case, the Kasner exponents as well as the Kasner axes are all changed. In contrast to the collinear case, the nature of singularities on $t = 0$ is anastigmatic. When $\alpha = 1/2$, we have $p_1 = 0$. Thus, the solutions with $\alpha = 1/2$ are free of spacetime singularities on the hypersurface $t = 0$ in the non-collinear case.

(ϵ) $\alpha > 1$. Then, the metric takes the form

$$ds^2 \simeq \varepsilon_0^{(4)} d\tau'^2 - \varepsilon_1^{(4)} \tau_1'^{2p'} dz^2 - \varepsilon_2^{(4)} \tau^{2p'_2} (dx^2)^2$$
$$- \varepsilon_3^{(4)} \tau'^{2p'_3} (dx^3)^2, \tag{4.133}$$

as $t \to 0^+$, where τ' and p_i are defined by Eqs. (4.131) and (4.132), respectively. Comparing this case with the last one, we find that in the present case only the Kasner exponents are changed, and that the Kasner axes remain unchanged relative to those in the collinear case. When $\alpha = 3/2$, we have $p_1 = 0$. That is, the solutions with $\alpha = 3/2$ are also free of spacetime curvature singularities on the hypersurface $t = 0$. The nature of the singularities is anastigmatic for $1 < \alpha < 3/2$ and astigmatic for $\alpha > 3/2$, whereas in the collinear case it is astigmatic for all the case with $\alpha > 1$.

To summarize the above results, we conclude that, if any of the following conditions holds:

$$\text{(i) } \alpha = -\frac{3}{2}, \quad \text{(ii) } \alpha = -\frac{1}{2}, \quad \text{(iii) } \alpha = \frac{1}{2}, \quad \text{or} \quad \text{(iv) } \alpha = \frac{3}{2}, \tag{4.134}$$

then the corresponding solutions are free of spacetime curvature singularities on the hypersurface $t = 0$. Therefore, these solutions are extendible across this surface. On the other hand, comparing Eqs. (4.134) and (4.127), we find that the solutions with $\alpha = -1/2$ or $-3/2$ are free of spacetime singularities on $t = 0$ in both of the collinear ($Q_1 = Q_2 = 0$) and non-collinear ($Q_1 Q_2 \neq 0$) cases. But, the solutions with $\alpha = 1/2$ or $3/2$ are free of spacetime singularities only in the non-collinear case. For the same reasons as those given in the one-soliton case, this is attributed to the interaction between different polarization modes.

Note that if we set the parameter Q_1 equal to zero, we shall rediscover the one-soliton solutions given by Eqs. (4.56)–(4.58) but with $\delta_{1,2}$ replaced by $\delta_{1,2} + 1/2$. Taking this fact into account, we find that the solutions with $\alpha = -3/2$ are free of spacetime curvature singularities on the hypersurface $t = 0$ only for the case where $Q_1 Q_2 \neq 0$, otherwise they will be all singular [see Eqs. (4.53), (4.76) and (4.134)].

4.3.3. Particular solutions

Let us now turn to some representative cases of the solutions given by Eqs. (4.110)–(4.111).

Case A: $\delta_1 = \delta_2 = 0$. In this case, we find that the solutions given by Eqs. (4.110)–(4.111) reduce to

$$\chi = \left[\left(1 - \eta^2 \right) \left(1 - \mu^2 \right) \right]^a \frac{A}{B},$$

$$q_2 = \left[\left(1 - \eta^2 \right) \left(1 - \mu^2 \right) \right]^{2a} \frac{C}{B}, \tag{4.135}$$

$$f = C_0 \frac{[(1 - \eta^2)(1 - \mu^2)]^{a(a-2) - 1/4}}{\eta^2 - \mu^2} A,$$

where the functions A, B and C are now defined by

$$
\begin{aligned}
A &\equiv (1 - \eta^2)(1 - \mu^2)^{2a}[(1 - \eta)^{2a} + Q_1 Q_2 (1 + \eta)^{2a}]^2 \\
&\quad + (1 - \mu^2)(1 - \eta^2)^{2a}[Q_1(1 + \mu)^{2a} - Q_2(1 - \mu)^{2a}]^2, \\
B &\equiv (1 - \mu^2)^{2a}[(1 + \eta)(1 - \eta)^{2a} - Q_1 Q_2(1 - \eta)(1 + \eta)^{2a}]^2 \\
&\quad + (1 - \eta^2)^{2a}[Q_1(1 - \mu)(1 + \mu)^{2a} + Q_2(1 + \mu)(1 - \mu)^{2a}]^2, \\
C &\equiv -2\{Q_1(\eta + \mu)[(1 + \mu)^{2a}(1 - \eta)^{2a} + Q_2^2(1 - \mu)^{2a}(1 + \eta)^{2a}] \\
&\quad + Q_2(\eta - \mu)[(1 - \mu)^{2a}(1 - \eta)^{2a} + Q_1^2(1 + \mu)^{2a}(1 + \eta)^{2a}]\}.
\end{aligned}
\tag{4.136}
$$

But, these are the solutions found by Ernst *et al.* (1987a, 1987b, 1988) by using a different method. When $\delta_1 = \delta_2 = 0$, Eq. (4.122) yields $n = m = 1$. Hence, from the previous analysis we can see that this subclass of solutions in general represents the collision of variably polarized gravitational shock + impulsive plane waves.

Case A.1: $a = 0$. If we further set $a = 0$, we can write the corresponding solutions in a simple form, by introducing the quantities p, q and l via the relations,

$$
Q_1 Q_2 = \frac{p - 1}{p + 1}, \quad Q_1 = \frac{l - q}{p + 1}, \quad Q_2 = \frac{l + q}{p + 1},
\tag{4.137}
$$

from which we obtain

$$
p^2 + q^2 = 1 + l^2.
\tag{4.138}
$$

Substituting Eq. (4.137) into Eqs. (4.135) and (4.136), and considering the fact that the parameter a vanishes in this case, we find that the corresponding solution takes the form,

$$
\begin{aligned}
\chi &= \frac{p^2 \left(1 - \eta^2\right) + q^2 \left(1 - \mu^2\right)}{(1 - p\eta)^2 + (l - q\mu)^2}, \\
q_2 &= \frac{2(q\mu - lp\eta)}{(1 - p\eta)^2 + (l - q\mu)^2}, \\
f &= C_0' \frac{p^2 \left(1 - \eta^2\right) - q^2 \left(1 - \mu^2\right)}{[(1 - \eta^2)(1 - \mu^2)]^{1/4}(\eta^2 - \mu^2)},
\end{aligned}
\tag{4.139}
$$

where $C_0' \left[= 4C_0(1 - p)^{-2}\right]$ is another arbitrary constant. The solution given by Eq. (4.139) is the Nutku–Halil solution with the NUT parameter l (Ernst, Garcia and Hauser, 1987b, 1988). Actually, it can be shown that the corresponding Ernst potential E introduced in Eq. (3.44) is now given by

$$E = \frac{1}{1 + l^2}[(p\eta + lq\mu) + i(q\mu - lp\eta)]. \tag{4.140}$$

Thus, Eq. (4.140) reduces to

$$E = p\eta + iq\mu, \tag{4.141}$$

when $l = 0$, which is the Ernst potential for the Nutku–Halil solution (Chandrasekhar and Ferrari, 1984).

It is interesting to note that the Kerr stationary axisymmetric solution (1963) also follows the same Ernst potential E given by Eq. (4.141) (Ernst, 1968a).

Case A.2: $a = 1/2$. Then, we find that the corresponding solution can be written in the form

$$\chi = \frac{(1 - p\eta)^2 + (l - q\mu)^2}{p^2(1 - \eta^2) + q^2(1 - \mu^2)}t,$$

$$q_2 = -2\frac{p(q - l\mu)(1 - \eta^2) - q(p - \eta)(1 - \mu^2)}{p^2(1 - \eta^2) + q^2(1 - \mu^2)},$$

$$f = C_0''\frac{(1 - p\eta)^2 + (l - q\mu)^2}{(\eta^2 - \mu^2)}, \tag{4.142}$$

where C_0'', p, q and l are now defined by (Economou and Tsoubelis, 1989),

$$\frac{Q_2}{Q_1} = \frac{q + l}{q - l}, \quad \frac{1}{Q_1} = \frac{1 + p}{q - l}, \quad Q_2 = \frac{1 - p}{q - l},$$

$$C_0'' = \frac{4C_0}{Q_1{}^2(q - l)^2}. \tag{4.143}$$

It is easy to prove that p, q and l defined by Eq. (4.143) also satisfy the relation (4.138). As first noticed by Ernst, Garcia and Hauser (1987b, 1988), the above solution is the Chandrasekhar–Xanthopoulos solution with the NUT parameter l. In fact, setting $l = 0$, we shall rediscover the Chandrasekhar–Xanthopoulos solution (Chandrasekhar and Xanthopoulos, 1986a). In this case, the corresponding stationary axisymmetric solution is the Kerr–NUT solution (Kramer *et al.*, 1980).

When $q = 0$, defining p' and q' as

$$p' = -p^{-1}, \quad q' = -lp^{-1}, \quad (q = 0),$$ (4.144)

we find that Eq. (4.142) reads

$$\chi = \frac{1 + 2p'\eta + \eta^2}{1 - \eta^2}t, \quad q_2 = 2q'\mu,$$

$$f = \tilde{C}''_0 \frac{1 + 2p'\eta + \eta^2}{\eta^2 - \mu^2}, \quad (q = 0),$$ (4.145)

where

$$\tilde{C}''_0 = \frac{4C_0}{Q_1^2 q'^2}, \quad p'^2 + q'^2 = 1,$$ (4.146)

which is the Ferrari–Ibañez solution (Ferrari and Ibañez, 1988). Thus, the Ferrari–Ibañez solution corresponds to the Taub–NUT solution (Hawking and Ellis, 1973).

Case B: $Q_1 = 0$ (or $Q_2 = 0$). Setting $Q_1 = 0$ in Eqs. (4.110)–(4.111), we find that the reduced solutions are the one-soliton solutions given by Eqs. (4.56)–(4.58) with the soliton parameters $\delta_{1,2}$ replaced by $\delta_{1,2} + 1/2$.

Similarly, if we set $Q_2 = 0$, instead of $Q_1 = 0$, in Eqs. (4.110)–(4.111) we find that the resulting solutions are also the one-soliton solutions, but now with the soliton parameters $\delta_{1,2}$ replaced by $\delta_{1,2} + 1/2$ and μ by $-\mu$.

Case C: $Q_1 \to \infty$ (or $Q_2 \to \infty$). If we set

$$C_0 = \frac{\overline{C}_0}{Q_1^2},$$ (4.147)

and then take the limit $Q_1 \to \infty$ (but keep $\overline{C}_0$ finite), we find that the solutions given by Eqs. (4.110)–(4.111) become

$$\chi = \frac{A}{B}\chi^{(0)},$$

$$q_2 = -2Q_2 \frac{(\eta - \mu)^{2\delta_- + 1}}{B}\left[\chi^{(0)}\right]^2,$$ (4.148)

$$f = \overline{C}_0 \frac{(1 - \eta)^{\rho_1}(1 - \mu)^{\rho_2}(1 + \eta)^{r'_1}(1 + \mu)^{r'_2}}{(\eta + \mu)^{(\delta_+ - 1)^2}(\eta - \mu)^{(\delta_- + 1)^2}}A,$$

but now the functions A and B are defined by

$$
\begin{aligned}
A \equiv \left(1 - \mu^2\right) & (1 - \eta)^{2(a+\delta_1)}(1 + \mu)^{2(a-\delta_2)} \\
&+ Q_2^2 \left(1 - \eta^2\right)(\eta - \mu)^{4\delta_-}(1 + \eta)^{2(a-\delta_1)}(1 - \mu)^{2(a+\delta_2)}, \\
B \equiv \left(1 - \mu^2\right)^2 & (1 - \eta)^{2(a+\delta_1)}(1 + \mu)^{2(a-\delta_2)} \\
&+ Q_2^2(1 - \eta)^2(\eta - \mu)^{4\delta_-}(1 + \eta)^{2(a-\delta_1)}(1 - \mu)^{2(a+\delta_2)},
\end{aligned}
\tag{4.149}
$$

and r'_A are given by

$$
r'_A = (a - \delta_A)^2 - \frac{1}{4}, \quad (A = 1, 2). \tag{4.150}
$$

Equations (4.120), (4.121) and (4.148) show that the appropriate choice of the parameters n and m now are

$$
n = [2 - (\delta_+ - 1)^2]^{-1}, \quad m = [2 - (\delta_- + 1)^2]^{-1}. \tag{4.151}
$$

Following the discussion given in Section 4.2, it can be shown that this subclass of solutions represents the same type of collision as the one-soliton solutions given by Eqs. (4.56)–(4.58) (Tsoubelis and Wang, 1992).

In a similar fashion, we find that in the limit $Q_2 \to \infty$, the two-soliton solutions become

$$
\chi = \frac{A}{B} \chi^{(0)},
$$

$$
q_2 = -2Q_1 \frac{(\eta + \mu)^{2\delta_+ + 1}}{B} \left[\chi^{(0)}\right]^2, \tag{4.152}
$$

$$
f = \overline{C}_0 \frac{(1 - \eta)^{\rho_1}(1 - \mu)^{\rho'_2}(1 + \eta)^{r'_1}(1 + \mu)^{r_2}}{(\eta + \mu)^{(\delta_+ + 1)^2}(\eta - \mu)^{(\delta_- - 1)^2}} A,
$$

where A, B and C are defined by

$$
\begin{aligned}
A \equiv \left(1 - \mu^2\right) & (1 - \eta)^{2(a+\delta_1)}(1 - \mu)^{2(a+\delta_2)} \\
&+ Q_1^2 \left(1 - \eta^2\right)(\eta + \mu)^{4\delta_+}(1 + \eta)^{2(a-\delta_1)}(1 + \mu)^{2(a-\delta_2)}, \\
B \equiv \left(1 - \mu^2\right)^2 & (1 - \eta)^{2(a+\delta_1)}(1 - \mu)^{2(a+\delta_2)} \\
&+ Q_1^2(1 - \eta)^2(\eta + \mu)^{4\delta} + (1 + \eta)^{2(a-\delta_1)}(1 + \mu)^{2(a-\delta_2)},
\end{aligned}
\tag{4.153}
$$

and $\overline{C}_0, r'_1$ and ρ'_2 are given by

$$
\overline{C}_0 \equiv C_0 Q_1^2, \quad r'_1 = (a - \delta_1)^2 - \frac{1}{4}, \quad \rho'_2 = (a + \delta_1)^2 - \frac{1}{4}. \tag{4.154}
$$

In this case, the appropriate choice of the parameters n and m is

$$n = [2 - (\delta_+ + 1)^2]^{-1}, \quad m = [2 - (\delta_- - 1)^2]^{-1}. \qquad (4.155)$$

Case D: $a = 0$. When $a = 0$, the corresponding solutions can be considered as the non-collinear generalization of the Szekeres family of colliding collinearly polarized gravitational plane waves, since if we further set $Q_1 = 0 = Q_2$ we shall rediscover the Szekeres solutions.

It must be noted that Halil "found" a class of non-diagonal solutions (1979) using the harmonic maps of Riemannian manifolds. As the author himself declared, it "is" a non-collinear generalization of the Szekeres family of solutions. However, Griffiths (1987) argued that the Halil solutions do not satisfy the Einstein vacuum equations.

In addition, Hassan, Feinstein and Manko (1990) developed an algorithm for constructing exact solutions of colliding plane gravitational waves. As an example, they found an Ernst potential E using the diagonal solutions given by Eqs. (4.23) and (4.24) as seed, but unfortunately they have not given the explicit solution for the function f. This is because in applying their algorithm one has to integrate the system (4.7a)–(4.7b) directly after the Ernst potential is given. However, for such a complicated Ernst potential this is not an easy task. It is expected that the specific Ernst potential given by Hassan *et al.* may correspond to the two-soliton solutions with $Q_1 = Q_2$.

Finally, we would like to mention that colliding axisymmetric pp waves were studied by Ivanov (1998), and Gürses, Kahya and Karasu (2002) generalized the collision of gravitational plane waves to high even-dimensional spacetimes, with the combinations of collinear and noncollinear polarized gravitational waves, and found that the spacetime singularity structure depends on the parameters of the solution.

Chapter 5

Collisions of Gravitational Waves with Matter Fields

In this chapter, we consider the collision of plane gravitational waves with matter. Specifically, in Section 5.1 two families of plane symmetric solutions of the Einstein field equations are presented. These solutions represent the collision of infinitely thin (impulsive) shells of null dust with the same kind of shells or with constantly (collinearly) or variably (non-collinearly) polarized gravitational plane waves. The general properties of these solutions are discussed. In Section 5.2, the collision of massless scalar waves is studied. A method for generating this kind of solutions from a known vacuum solution is presented. Some relevant topics are discussed. Following it, in Section 5.3 we consider solutions that represent the collision of oppositely moving "null dust" clouds. With specific examples, the interesting questions about the uniqueness of the outcome of the above kind of collisions and the gravitational "phase transitions" induced from such collisions are investigated. By solving the corresponding Einstein–Maxwell–Weyl field equations, the ambiguity is resolved and the gravitational "phase transitions" from massless particles into massive particles is not allowed in general relativity.

In Section 5.4, we consider the collision of two electromagnetic plane waves. In this section, we mainly focus on the well-known Bell–Szekeres solution (1974), as it already contains the main features of this type of collisions, although it is quite simple and represents the collision of two pure collinearly polarized electromagnetic shock plane waves. In particular, the focusing surface is a Cauchy horizon, instead of a spacetime curvature singularity, and its extension beyond this surface is not unique (Clarke and Hayward, 1989), but it is expected that such a horizon is not stable, and in more realistic case, it should be replaced by a spacetime singularity (Chandrasekhar and Xanthopoulos, 1988; Konkowski and Helliwell, 1991).

Finally, in Section 5.5 the collisions of other matter waves in Einstein's theory as well as in other theories of gravity, including string/M-theory, are briefly discussed. A very brief comment on the collisions of gravitational waves in curved backgrounds, such as FRW and AdS, is also provided.

5.1. Collision of Impulsive Shells of Null Dust and Gravitational Plane Waves

The infinitely thin (or impulsive) shells have attracted a lot of attention since Israel's work (Israel, 1966, 1967) was published (Boulware, 1973; Bronnikov, 1980; Siegel, 1981; Dray and 't Hooft, 1985a, 1985b; Tsoubelis, 1989b; Barrabés and Hogan, 2003; Bronnikov, Santos and Wang, 2019; and references therein). One of the main reasons is that impulsive shells can be considered as good mathematical models for describing dust clouds analogous to the models used for impulsive gravitational waves. However, the interaction of two such shells had not been studied until 1986 when Dray and 't Hooft first discussed the gravitational effects of colliding planar shells of matter. It was shown that, quite similar to the Khan–Penrose solution of two colliding impulsive gravitational waves (Khan and Penrose, 1971), such a collision develops a curvature singularity, and that a Coulomb-like gravitational field appears after the collision. The above results together with the ones obtained from the study of the effect of colliding null dust clouds on the formation of singularities (Chandrasekhar and Xanthopoulos, 1987a) led us once again to believe that the collision of impulsive shells always produces curvature singularities.

However, in 1991 Tsoubelis and Wang found a two-parameter class of solutions of the Einstein field equations (Tsoubelis and Wang, 1991) and found that, like the Chandrasekhar–Xanthopoulos vacuum solution (Chandrasekhar and Xanthopoulos, 1986a), the collision of such shells does not inevitably develop curvature singularities, and that, in contrast to the vacuum case (Szekeres, 1972), the Coulomb-like gravitational field does not necessarily appear in the interaction region.

5.1.1. Colliding impulsive shells with collinearly polarized gravitational waves

In this subsection, let us consider the two-parameter class of solutions found by Tsoubelis and Wang (1991) in detail. As mentioned above, this class of solutions represents the collision of null dust shells and collinearly polarized gravitational plane waves.

The analysis given in Section 3.2 [especially, see Eqs. (3.25) and (3.27)] shows that, for any vacuum solution given in the interaction region (Region IV), the extended solution obtained by the Khan–Penrose substitutions (3.11) will satisfy the Einstein vacuum equations inside the pre-collision regions (Regions I–III). But across the hypersurfaces $u = 0$ and $v = 0$ the Ricci tensor will suffer the following discontinuities:

$$R_{\mu\nu} = \frac{2nu^{2n-1}}{1 - v^{2m}H(v)}\delta(u)\hat{l}_\mu\hat{l}_\nu + \frac{2mv^{2m-1}}{1 - u^{2n}H(u)}\delta(v)\tilde{n}_\mu\hat{n}_\nu, \qquad (5.1)$$

where

$$\hat{l}_\mu \equiv Al_\mu = \delta^u_\mu, \quad \hat{n}_\mu \equiv Bn_\mu = \delta^v_\mu. \qquad (5.2)$$

Using the Einstein field equations given by Eq. (1.20), we immediately obtain

$$T_{\mu\nu} = \frac{2nu^{2n-1}}{1 - v^{2m}H(v)}\delta(u)\hat{l}_\mu\hat{l}_\nu + \frac{2mv^{2m-1}}{1 - u^{2n}H(u)}\delta(v)\hat{n}_\mu\hat{n}_\nu. \qquad (5.3)$$

It is clear that when $n = 1/2$ the first term on the right-hand side of Eq. (5.3) represents an impulsive shell of null dust with support on the hypersurface $u = 0$, while when $m = 1/2$ the second term represents a similar shell but with support on the hypersurface $v = 0$. Therefore, when $n = 1/2$ and $m = 1/2$ the solutions given by Eqs. (4.23) and (4.24) will represent the collision of two impulsive shells of null dust, each of which may be accompanied by a constantly polarized gravitational plane wave. When one of the two parameters n and m is equal to $1/2$ and the other one still satisfies the conditions (4.29), the above solutions will represent the collision of an impulsive shell of null dust with a gravitational plane wave.

Without loss of generality, let us consider the solutions given by Eqs. (4.23) and (4.24) for the case where

$$\text{(i) } (n, m) = \left(\frac{1}{2}, \frac{1}{2}\right), \quad \text{or} \quad \text{(ii) } (n, m) = \left(\frac{1}{2}, \geq 1\right), \qquad (5.4)$$

which implies that in all the cases to be considered there always exists an impulsive shell of null dust that propagates along the hypersurface $u = 0$ toward the left in Fig. 3.1. From Eq. (4.28) we can see that $n = 1/2$ implies

$\delta_+ = 0$, or

$$\delta_1 = -\delta_2, \left(n = \frac{1}{2}\right). \tag{5.5}$$

Then, the combination of Eqs. (3.24), (4.23), (4.24) and (5.4) leads to the following expressions for the non-vanishing Weyl scalars in Region IV,

$$\Psi_0^{IV}(u,v) = 8m^2 v^{m-2} \left[\frac{b_1(av^{3m} + \delta_1 Y^3)}{t^2} \right.$$
$$\left. - \frac{a\delta_1[(2a + \delta_1)v^m + (a + 2\delta_1)Y]}{(v^m + Y)^2} \right],$$

$$\Psi_2^{IV}(u,v) = \frac{2b_1 m v^{2m-1}}{t^2} + \frac{2a\delta_1 m v^{m-1}}{Y(v^m + Y)^2}, \tag{5.6}$$

$$\Psi_4^{IV}(u,v) = \frac{2b_1(\delta_1 v^{3m} + aY^3)}{Y^3 t^2}$$
$$- \frac{2a\delta_1[(2a + \delta_1)Y + (a + 2\delta_1)v^m]}{Y^3(v^m + Y)^2}.$$

From Eq. (5.6) we can see that as $t \to 0^+$ (while $u, v \neq 0$), a space-time curvature singularity develops in all of the models, except for the case in which Eq. (4.46) holds. In the latter, one of the Killing vectors ∂_x and ∂_y becomes null as $t \to 0^+$, as can be seen from Eqs. (4.23), (4.24) and (5.4). Thus, when Eq. (4.46) holds, the space-like singularity that otherwise is formed on the hypersurface $t = 0$ gives its place to a Cauchy horizon, beyond which the metric can be extended, although such extensions are not unique. In addition, following Yurtsever's arguments (1987) and Griffiths' proof for pure gravitational waves, these horizons are expected not stable, too.

On the other hand, from Eqs. (3.24) and (5.6), it can be shown that in Region II the only non-vanishing Weyl scalar is Ψ_0, while in Region III the only non-vanishing one is Ψ_4, which are given, respectively, by

$$\Psi_0^{II}(v) = \frac{2m}{(1 - v^{2m})^2}\{am(4a^2 - 1)v^{4m-2} + 12\delta_1 a^2 m v^{3m-2}$$
$$+ 3a(2m - 1)v^{2m-2} + \delta_1(m - 1)v^{2m-2}\}, \tag{5.7}$$

$$\Psi_4^{III}(u) = \frac{a(4a^2 - 1)}{2(1 - u)^2}. \tag{5.8}$$

In Region I, all the Weyl scalars vanish.

Combining Eqs. (3.24), (3.26) with Eqs. (5.6)–(5.8), we find that the non-vanishing Weyl scalars have the following behavior when across the hypersurface $u = 0$ or $v = 0$.

I $\rightarrow$ III:

$$\Psi_0, \Psi_2 \text{ are continuous,}$$
$$\Psi_4^{\text{I--III}} = \Psi_4^{\text{III}}(u = 0)H(u) + a\delta(u). \tag{5.9}$$

II $\rightarrow$ IV:

$$\Psi_0 \text{ is continuous,}$$
$$\Psi_2^{\text{II--IV}} = \Psi_2^{\text{IV}}(u = 0, v)H(u), \tag{5.10}$$
$$\Psi_4^{\text{II--IV}} = \Psi_4^{\text{IV}}(u = 0, v)H(u) + \frac{(a + kv^m)}{X^2}\delta(u).$$

I $\rightarrow$ II:

$$\Psi_2, \Psi_4 \text{ are continuous,}$$
$$\Psi_0^{\text{I--II}} = \Psi_0^{\text{II}}(v = 0)H(v) + 2m(kv^{m-1} + av^{2m-1})\delta(v). \tag{5.11}$$

III $\rightarrow$ IV:

$$\Psi_4 \text{ is continuous,}$$
$$\Psi_0^{\text{III--IV}} = \Psi_0^{\text{IV}}(u, v = 0)H(v) + \frac{2m(kv^{m-1}Y + av^{2m-1})}{Y^2}\delta(v), \tag{5.12}$$
$$\Psi_2^{\text{III--IV}} = \Psi_2^{\text{IV}}(u, v = 0)H(v),$$

where

$$k \equiv \delta_1 = -\delta_2. \tag{5.13}$$

To study the above solutions further, let us consider some typical cases.

Case A: $m = 1/2$. In this case, from Eqs. (4.28) and (5.4) we find

$$k = 0, \tag{5.14}$$

for which we find $m = 1/2$, as it can be seen form Eq. (4.28). Then, setting $n = 1/2$ and $m = 1/2$ in Eq. (5.3), we find that the non-vanishing components of $T_{\mu\nu}$ are given by

$$T_{uu} = \frac{\delta(u)}{1 - vH(v)}, \quad T_{vv} = \frac{\delta(v)}{1 - uH(u)}, \tag{5.15}$$

which correspond to two impulsive shells of null dust with support, respectively, on $u = 0$ and $v = 0$. On the other hand, from Eqs. (5.6)–(5.14) we find,

$$\Psi_0 = \frac{a(4a^2 - 1)H(v)}{2(1 - uH(u) - v)} + \frac{a\delta(v)}{1 - uH(u)}, \qquad (5.16)$$

$$\Psi_4 = \frac{a(4a^2 - 1)H(u)}{2[1 - u - vH(v)]} + \frac{a\delta(u)}{1 - vH(v)}, \qquad (5.17)$$

$$\Psi_2 = \frac{(4a^2 - 1)H(u)H(v)}{4[1 - u - v]^2}, \qquad (5.18)$$

which show that the corresponding solutions represent the symmetric collision of a pair of impulsive shells of null dust, each of which is in general accompanied by a constantly polarized impulsive + shock gravitational plane wave. As a result of the collision, a Coulomb-like gravitational field Ψ_2 appears in the interaction region (Region IV) and a spacetime singularity is finally developed on the hypersurface $1 = u + v$.

When $a = 0$, Ψ_0 and Ψ_4 vanish everywhere, and the corresponding solution represents the collision of two pure impulsive shells of null dust. This solution was first found by Dray and 't Hooft (1986) and shown by Tsoubelis (1989b) that it belongs to the Szekeres family of solutions (Szekeres, 1972).

When $a = \pm 1/2$, Eqs. (5.16) and (5.17) show that the two incoming gravitational waves are impulsive ones. Thus, in this case the solutions represent the collision of two impulsive shells of null dust, each of which is accompanied by an impulsive gravitational plane waves. One of the remarkable features in this case is that the interaction region (Region IV) is flat and no Coulomb-like gravitational field appears in this region. Comparing it, for example, with the Khan–Penrose solution (Khan and Penrose, 1971), we are led to the conclusion that, when accompanied by null dust shells, the collision of two impulsive gravitational waves does not necessarily give rise to a Coulomb-like gravitational field in the interaction region. Since in the present case Region IV is flat, no spacetime singularity develops on the hypersurface $1 = u + v$.

Before turning to the next case, let us note that the solutions presented in this case were first obtained by Stoyanov (1979), but with an incorrect interpretation that the above solutions represented the collision of two purely gravitational waves. The fact that Stoyanov's interpretation cannot be supported was first noticed by Nutku (1981) and is made clear from the analysis presented above.

Case B: $m = 1$. When $m = 1$, we find $k = \pm 1/2$ and

$$T_{uu} = \frac{\delta(u)}{1 - v^2 H(v)}, \quad T_{vv} = 0, \tag{5.19}$$

and

$$\Psi_4^{\mathrm{I-III}} = \frac{1}{2} a(4a^2 - 1)H(u) + a\delta(u), \tag{5.20}$$

$$\Psi_0^{\mathrm{I-II}} = 6aH(v) + 2k\delta(v), \tag{5.21}$$

$$\Psi_4^{\mathrm{II-IV}} = \frac{a[3v^2 + 12akv + (4a^2 - 1)]}{2(1 - v^2)^2} H(u) + \frac{a + kv}{1 - v^2}\delta(u), \tag{5.22}$$

$$\Psi_0^{\mathrm{III-IV}} = \frac{6a}{1 - u}H(v) + \frac{2k}{\sqrt{1 - u}}\delta(v). \tag{5.23}$$

Equation (5.19) shows that in this class of solutions, the hypersurface $v = 0$ is always free of matter, while Eqs. (5.20) and (5.21), on the other hand, show that the solutions with $a \neq 0, \pm 1/2$ represent the collision of an impulsive shell of null dust with an impulsive + shock gravitational wave. The former is accompanied by an impulsive + shock gravitational wave. In the present case, all the solutions are singular on the hypersurface $1 - u - v^2 = 0$, except for the ones in which $a = 0, \pm 1$ [see Eq. (4.46)].

Case B.1: $a = 0$. In this case, Eqs. (5.20) and (5.21) reduce to $\Psi_4^{\mathrm{I-III}} = 0$ and $\Psi_0^{\mathrm{I-II}} = \epsilon\delta(v)$, respectively, where $\epsilon = \pm 1$. In fact, in this case the only non-vanishing Weyl scalars are

$$\Psi_0(u, v) = \frac{\epsilon}{\sqrt{1 - uH(u)}}\delta(v), \quad \Psi_4(u, v) = \frac{\epsilon v H(v)}{2(1 - v^2)}\delta(u). \tag{5.24}$$

Therefore, this model represents the collision of an impulsive gravitational plane wave incident from the left-hand side along the null hypersurface $v = 0$ with an impulsive shell of null dust incident from the right-hand side along the null hypersurface $u = 0$. It is very interesting to note that an impulsive gravitation wave represented by Ψ_4 along the null hypersurface $u = 0$ is produced right after the collision. The spacetime is flat inside all the four regions (Regions I–IV) and the only effect of the collision is the mutual focusing of the wave pulse and the null dust shell, and isolated spacetime singularities are developed, respectively at the points $(u, v) = (0, 1)$ and $(u, v) = (1, 0)$. The corresponding metric was first obtained by Babala (1987) and rediscovered by Tsoubelis (1989b).

Case B.2: $a = \pm 1/2 \equiv \epsilon'/2$. In this case, Eqs. (5.20) and (5.21) reduce to

$$\Psi_4^{\text{I–III}} = \frac{1}{2}\epsilon'\delta(u), \tag{5.25}$$

$$\Psi_0^{\text{I–II}} = 3\epsilon' H(v) + \epsilon\delta(v), \tag{5.26}$$

where ϵ and ϵ' can be chosen independently, always we have $|\epsilon| = |\epsilon'| = 1$. Then, we can see that the present model represents the collision of a shock wave with an impulsive shell of null dust, each of which is accompanied by an impulsive gravitational wave. In this case, the formation of a spacetime singularity along the hypersurface $u = 1 - v^2$ is inevitable.

Case B.3: $a = -2k$. Since in this case we have

$$\Psi_4^{\text{I–III}} = -\frac{3}{2}\varepsilon H(u) - \varepsilon\delta(u), \tag{5.27}$$

$$\Psi_0^{\text{I–II}} = -6\varepsilon H(v) + \varepsilon\delta(v), \tag{5.28}$$

we find that the corresponding solutions represent the collision of two impulsive + shock gravitational waves, one of which is accompanied by an impulsive shell of null dust. The collision is such that the spacetime is free of singularity on the hypersurface $1 - u - v^2 = 0$. In fact, the Coulomb-like gravitational field that develops in Region IV after the collision remains finite as one approaches the hypersurface $t = 1 - u - v^2 = 0$, and is given by

$$\Psi_2^{\text{IV}}(u, v) = -\frac{1}{[\sqrt{1 - u} + v]^2\sqrt{1 - u}}. \tag{5.29}$$

Thus, we have

$$\Psi_2^{\text{IV}}(u = 1 - v^2, v) = -\frac{1}{4v^3}. \tag{5.30}$$

Case C: $m = 2$. In this case, we have $k = \epsilon\sqrt{3}/8$ and that

$$\Psi_4^{\text{I–III}} = \frac{1}{2}a(4a^2 - 1)H(u) - a\delta(u), \tag{5.31}$$

$$\Psi_0^{\text{I–II}} = 4kH(v). \tag{5.32}$$

Thus, in the present case the solutions represent the collision of a shock gravitational wave incident from the left with an impulsive shell of null dust incident from the right. The latter is accompanied by a gravitational shock + impulsive wave, provided $a \neq 0, \pm 1/2$. The behavior of this subclass of models after collision is similar to the ones obtained in Cases A and B above. Specifically, the development of a spacetime singularity along the $1 - u - v^4 = 0$ is inevitable, except for the cases where $a = -k \pm 1/2$.

When $a = 0$, one of the two incoming impulsive shells of null dust appearing in the Dray–'t Hooft model is replaced by a shock gravitational plane wave. The outcome of the collision is similar to that in the Dray–'t Hooft model. Specifically, a Coulomb-like gravitational field appears in the interaction region, and a spacetime singularity develops on the hypersurface $t = 1 - u - v^4 = 0$.

When $a = \pm 1/2$, the null dust shell is accompanied by an impulsive gravitational wave. Thus, the corresponding solutions represent the collision of a shock gravitational wave and an impulsive shell of null dust which is accompanied by an impulsive gravitational wave. In this case, a spacetime singularity also always develops on the hypersurface $t = 1 - u - v^4 = 0$.

Case D: $m = 4$. The shock wave incident from the left in Case C now is replaced by a gravitational plane wave with a smooth wavefront. This follows from the fact that in this case we have

$$\Psi_0^{\text{I–III}} = \Psi_0^{\text{III–IV}} = 0. \tag{5.33}$$

On the other hand, we have

$$\Psi_4^{\text{I–II}} = \frac{1}{2}a(4a^2 - 1)H(u) - a\delta(u), \tag{5.34}$$

as in the previous cases, and

$$\Psi_4^{\text{II–IV}} = \frac{a(4a^2 - 1) + 12a^2kv^4 + 12ak^2v^8 + k(4k^2 - 1)v^{12}}{2(1 - v^8)^2}H(u)$$

$$+ \frac{a + kv^4}{1 - v^8}\delta(u). \tag{5.35}$$

Comparing Eqs. (5.33) and (5.34) with Eqs. (5.32) and (5.31), we find that the current models represent the same kind of collisions as the $m = 2$ models, except that the shock gravitational wave incident in Region II is now replaced by a gravitational wave with a smooth wavefront. Thus, in the following we do not go to detail, except for mentioning that in the present case the second gravitational wave production is illustrated most clearly, although it arises essentially in all the spacetime models constructed above. Consider in this direction, the $a = 0$ case. For this particular model $\Psi_4^{\text{I–III}} = 0$. Therefore, the null dust shell incident from the right is not accompanied by any gravitational radiation. According to Eq. (5.35), on the other hand, we have

$$\Psi_4^{\mathrm{II-IV}} = \frac{3kv^{12}}{8(1-v^8)^2}H(u) + \frac{kv^4}{1-v^8}\delta(u), \qquad (5.36)$$

which shows that, upon interacting with the gravitational wave pulse incident from the left, the shell of null dust stimulates the emission of gravitational radiation in its own direction, besides getting focused as it is propagating.

5.1.2. Collisions of an impulsive null dust shell and a non-collinearly polarized gravitational wave

The analysis carried out in the last section shows that the one-soliton solutions given by Eqs. (4.56)–(4.58) will in general represent the collision of an impulsive shell of null dust with a variably polarized gravitational plane wave, if the soliton parameters δ_1 and δ_2 are chosen so that

$$\delta_+ \equiv \delta_1 + \delta_2 = 0, \qquad (5.37)$$

which leads to

$$n = \frac{1}{2}, \qquad (5.38)$$

as follows from Eq. (4.61). Note that the solutions given by Eqs. (4.56)–(4.58) are for the case $\delta_- \geq 0$. Thus, in this case we always have $m \geq 1$ [see Eq. (4.61)]. That is, the hypersurface $v = 0$ in this case is always free of matter.

From Eqs. (4.63) and (4.64), on the other hand, we can see that the solutions in Region II do not depend on δ_+. Consequently, the condition given by Eq. (5.37) does not have any restriction on the solutions given by Eqs. (4.63) and (4.64). Therefore, the gravitational plane wave incident in Region II is the one described below Eq. (4.65).

The plane gravitational wave incident in Region III can be obtained from Eqs. (5.8) and (5.9) [see Eqs. (4.33) and (4.34)] and is given by

$$\Psi_4(u, v < 0) = \frac{1}{2}a[(4a^2 - 1)Y^{-4}H(u) + 2\delta(u)]. \qquad (5.39)$$

It is clear from Eqs. (5.3) and (5.39) that, when $a \neq 0, \pm 1/2$, the impulsive shell of null dust with support on the hypersurface $u = 0$ is accompanied generally by a constantly polarized gravitational impulsive + shock plane wave. When $a = 1/2$, the shock part of the latter disappears and only the impulsive gravitational plane wave remains.

As an example, let us consider the solutions of Eqs. (4.56)–(4.58) with the parameters δ_1 and δ_2 set equal to zero, for which Eq. (4.61) yields

$$n = \frac{1}{2}, \quad m = 1. \tag{5.40}$$

The previous analysis shows that this class of solutions in general represents the collision of a variably polarized shock + impulsive gravitational plane wave with an impulsive shell of null dust. The latter may be accompanied by a constantly polarized shock + impulsive gravitational wave. The corresponding solutions are given by

$$\chi = \frac{A}{B} t^{2a+1}, \quad q_2 = -4qvY\frac{t^{4a}}{B}H(v),$$

$$e^{-M} = \frac{A}{Y} t^{2a(a-1)}, \quad e^{-U} = t = 1 - uH(u) - v^2 H(v), \tag{5.41}$$

where

$$A = [Y - vH(v)]^{4a} + q^2[Y + vH(v)]^{4a},$$

$$B = [Y - vH(v)]^{4a}[Y + vH(v)]^2 \tag{5.42}$$

$$+ q^2[Y + vH(v)]^{4a}[Y - vH(v)]^2,$$

$$Y = [1 - uH(u)]^{1/2}.$$

From the expressions of Eqs. (5.41) and (5.42) it can be seen that to study this class of solutions as a whole is very complicated. In the following, we consider only several representative cases, which are sufficient for illustrating the main properties of this class of solutions.

Case 1: $a = 0$. In this case, we find

$$\Psi_i^{\mathrm{IV}} = 0 \quad (i = 0, 2, 4), \tag{5.43}$$

which means that the spacetime is flat in the interaction region (Region IV). Hence, from Eqs. (3.24) and (3.26) we find that

$$\Psi_0(u,v) = \frac{1}{(1+q^2)Y}\{(1-q^2) - i2q\}\delta(v),$$

$$\Psi_4(u,v) = \frac{vH(v)}{2(1-v^2)D}\{(1-q^2)(1-v^2) + i2q(1+v^2)\}\delta(u), \quad (5.44)$$

$$\Psi_2(u,v) = 0,$$

where

$$D \equiv [(1+q^2)^2(1-v^2H(v))^2 + 16q^2v^2H(v)]^{1/2}. \tag{5.45}$$

Thus, in this case the solution represents the collision of an impulsive gravitational plane wave and an impulsive shell of null dust. Since the spacetime is flat in Region IV, the hypersurface $t = 0$ is free of spacetime curvature singularity, except at the two focusing points $(u, v) = (0, 1)$ and $(u, v) = (1, 0)$.

The study of polarization of the two impulsive gravitational waves given by Eq. (5.44) in the next subsection will reveal that this solution is actually the Babala solution referred as Case B.1 in Section 5.1.

Case 2: $a = 1/2$. In this case, if we define the functions $F(x, y)$ and $G(x, y)$ as

$$
\begin{aligned}
F(x, y) &\equiv (x - y)^5 - 6q^2 x(x^2 - y^2)^2 + q^4(x + y)^5 + i2q(x^2 - y^2) \\
&\quad \times [(1 + q^2)(5x^2 + y^2)y - 2(1 - q^2)(x^2 + 2y^2)x], \\
G(x, y) &\equiv (x - y)^4(x^2 + xy + y^2) - q^4(x + y)^4(x^2 - xy + y^2) \\
&\quad - 6q^2 xy(x^2 - y^2)^2 - i2q(1 + q^2)(x^2 - y^2)^3,
\end{aligned}
\tag{5.46}
$$

we find that the non-vanishing Weyl scalars in Region IV are given by

$$
\begin{aligned}
\Psi_0^{IV}(u, v) &= \frac{3(1 + q^2)Y}{tA^2 D} F(Y, v), \\
\Psi_4^{IV}(u, v) &= -\frac{3(1 + q^2)v}{4tA^2 DY^2} \bar{F}(v, Y), \\
\Psi_2^{IV}(u, v) &= \frac{1}{2t^2 A^2 Y} G(Y, v),
\end{aligned}
\tag{5.47}
$$

but now the function D is defined by

$$
D \equiv [A^2 + 16q^2 v^2 Y^2]^{1/2}.
\tag{5.48}
$$

Then, from Eqs. (3.24) and (3.26) we find

$$
\begin{aligned}
\Psi_0(u, v) &= \Psi_0^{IV}(uH(u), v)H(v) + \frac{1}{(1 + q^2)Y}\{(1 - q^2) - i2q\}\delta(v), \\
\Psi_4(u, v) &= \Psi_0^{IV}(u, v)H(u)H(v) + \frac{1 + q^2}{2AD}\{(1 - vH(v))^3 \\
&\quad + q^2(1 + vH(v))^3 + i2qv(1 - v^2)H(v)\}\delta(u), \\
\Psi_2(u, v) &= \Psi_2^{IV}(u, v)H(u)H(v).
\end{aligned}
\tag{5.49}
$$

Taking Eq. (5.38) into account, we find that this solution represents the collision of a variably polarized gravitational shock + impulsive wave with an impulsive shell of null dust, which is accompanied by a constantly polarized gravitational impulsive wave. In this model, Region III is flat, while Region II is curved due to the presence of the variably polarized gravitational wave.

As $t \to 0^+$, all the non-vanishing Weyl scalars become unbounded. Thus, the spacetime now is singular on the hypersurface $t = 0$.

Case 3: $a = 1$. Then, we find that the non-vanishing Weyl scalars are given by

$$\Psi_0(u, v) = \Psi_0^{IV}(uH(u), v)H(v) + \frac{1}{(1 + q^2)Y}\{(1 - q^2) - i2q\}\delta(v),$$

$$\begin{aligned}
\Psi_4(u, v) = &\Psi_4^{IV}(u, vH(v))H(u) + \frac{1}{2tAD}\{(2 + vH(v))(1 - vH(v))^8 \\
&+ q^2(1 + vH(v))^3[q^2(2 - vH(v))(1 + vH(v))^5 \\
&+ 4(1 + 3v^2H(v))(1 - vH(v))^3] \\
&+ i2qv(1 - v^2)H(v)[(1 - v)^4(1 + 4v + v^2) \\
&+ q^4(1 + v)^4(1 - 4v + v^2)]\}\delta(u),
\end{aligned} \tag{5.50}$$

$$\Psi_2(u, v) = \Psi_2^{IV}(u, v)H(u)H(v),$$

where

$$\Psi_0^{IV}(u, v) = \frac{6}{A^2D}F(Y, v), \quad \Psi_4^{IV}(u, v) = \frac{3}{2Y^2A^2D}F(v, Y),$$

$$\Psi_2^{IV}(u, v) = \frac{6}{A^2Y}G(Y, v), \tag{5.51}$$

and the functions D, F and G now are defined by

$$D(u, v) \equiv [A^2 + 16q^2v^2t^2Y^2]^{1/2},$$

$$\begin{aligned}
F(x, y) \equiv &(x - y)^{10} - q^2(x^2 - y^2)(x - y)^4(5x^4 + 30x^3y + 8x^2y^2 \\
&+ 2xy^3 + 3y^4) - q^4(x^2 - y^2)(x + y)^4(5x^4 - 30x^3y \\
&+ 8x^2y^2 - 2xy^3 + 3y^4) + q^6(x + y)^{10} \\
&+ i4qx[(x - y)^7(y^2 - 2xy - x^2) + 2q^2y(x - y)^3(5x^2 + y^2) \\
&- q^4(x + y)^7(y^2 + 2xy - x^2)],
\end{aligned} \tag{5.52}$$

$$\begin{aligned}
G(x, y) \equiv &(x - y)^6 - 12q^2xy(x^2 - y^2)^2 - q^4(x + y)^6 \\
&- i2q[(x - y)^4(y^2 + 4xy + x^2) \\
&+ q^2(x + y)^4(y^2 - 4xy + x^2)].
\end{aligned}$$

From Eqs. (5.3), (4.39) and (5.50), we can see that the solution in the present case represents the collision of a variably polarized gravitational impulsive + shock wave with the same type wave but constantly polarized. The latter has an impulsive shell of null dust as its leading front.

As $t \rightarrow 0^+$, Eqs. (5.51) and (5.52) show that all the non-vanishing Weyl scalars keep finite provided $q \neq 0$. That is, the spacetime is free of spacetime curvature singularity on the hypersurface $t = 0$ when $q \neq 0$. However, when $q = 0$, it is easy to see that the non-vanishing Weyl scalars become unbounded as $t \rightarrow 0^+$. Thus, in the latter case the spacetime is singular on $t = 0$. Consequently, a close relation is established between the collision involving impulsive shells of null dust, on the one hand, and the collision involving only gravitational waves, on the other. For the latter type of collision, the effect of polarization of colliding plane gravitational waves on the formation of singularities was already studied in Sections 4.2 and 4.3.

Before proceeding to the next section, we note that solutions that represent the collision of an impulsive shell of null dust and a variably polarized gravitational plane wave were first studied by Feinstein and Senovilla (1989). Starting with the assumption that the non-diagonal term q_2 (in their notation it is denoted by ω) depends only on a null coordinate v in the interaction region, they found a unique solution of the Einstein vacuum equations for a given arbitrary $q_2(v)$. After extending the solutions back to the pre-collision regions, they found an impulsive shell of null dust to appear on the hypersurface $u = 0$. It was the latter that enabled them to interpret their solutions as representing the collision of a plane gravitational wave with an impulsive shell of null dust.

5.1.3. Polarizations of colliding plane gravitational waves when coupled with an impulsive null dust shell

As we showed in Section 3.4, due to the interaction between an impulsive gravitational wave and the same kind of shells, the polarization of the impulsive gravitational wave gets rotated, while the polarization of a shock gravitational plane wave can be changed by both the nonlinear interaction with the other plane gravitational wave and the interaction with matter fields.

In this section, we consider the first two solutions discussed in the last subsection to illustrate further the polarizations of gravitational impulsive waves as well as the shock ones. The main reasons that in this subsection we do not include the study of the third solution discussed in the last subsection are two-fold. First, the expressions of the non-vanishing Weyl scalars for this solution are very complicated. It is difficult to see the physics. Second, from this solution we shall obtain the similar results about the polarizations of colliding plane gravitational waves as we shall get from the first two.

With the above in mind, let us first consider Case 1 solution. The non-vanishing Weyl scalars in this case are given by Eq. (5.44). Since on the hypersurface $v = 0$, we have $W = 0$ [see Eq. (5.41)]. Hence, integrating Eq. (3.61) we find that

$$\varphi_0^0 = 0. \tag{5.53}$$

Combining Eqs. (2.46), (5.44) and (5.53), we find that

$$\theta_0^{\mathrm{Im}} = \frac{1}{2} \tan^{-1} \left(\frac{2q}{1 - q^2} \right). \tag{5.54}$$

Obviously, $\theta_{0,u}^{\mathrm{Im}} = 0$. That is, the polarization angle of the impulsive gravitational wave represented by Ψ_0 does not change, since in the present case Ψ_0^{Im} does not interact with any of the other components of the gravitational field [see Eq. (3.69)].

On the other hand, integrating Eq. (3.59), we obtain the following solution:

$$\varphi_4^{(0)} = \frac{1}{2} \tan^{-1} \left(\frac{4q(1 - q^2)v^2}{(1 + q^2)^2 - (1 - 6q^2 + q^4)v^2} \right) H(v). \tag{5.55}$$

From Eqs. (2.38), (5.44) and (5.55), we find

$$\theta_4^{\mathrm{Im}} = \frac{1}{2} \tan^{-1} \left(\frac{2q}{1 - q^2} \right) H(v). \tag{5.56}$$

It follows that after it is created from the collision, the impulsive gravitational wave, represented by Ψ_0^{Im}, does not change its polarization, too. On the other hand, from Eqs. (5.54) and (5.56) we can see that the impulsive plane gravitational wave, Ψ_4, has the same polarization angle as the Ψ_0-wave does.

The above observations show that if we make a rotation in the (x^2, x^3)-plane, the metric should be brought into a diagonal form. In fact, we find that the metric takes the form

$$ds^2 = \frac{2(1 + q^2)}{Y} du\, dv - [Y - vH(v)]^2 (dx'^2)^2 - [Y + vH(v)]^2 (dx'^3)^2, \tag{5.57}$$

after making a coordinate rotation given by Eq. (2.32) with the angle given by Eq. (5.54). Equation (5.57) is the solution first found by Babala (1987).

Actually, if we replace u by u' defined as

$$u' = \begin{cases} 1 - 2(1-u)^{1/2}, & u \geq 0, \\ u - 1, & u \leq 0, \end{cases} \tag{5.58}$$

Eq. (5.57) will take the exact form used by Babala (1987).

Now let us turn to the solution of Case 2 given in Subsection 5.1.2. We first note that in this case the Weyl scalars Ψ_0 and Ψ_4 consist of two parts: the shock part and the impulsive part. In the following, we consider them, separately.

Let us first consider the impulsive part. Following the discussion given in the last case, we find that

$$\begin{aligned} \theta_0^{\text{Im}} &= \frac{1}{2}\tan^{-1}\left(\frac{2q}{1-q^2}\right), \\ \theta_4^{\text{Im}} &= \frac{1}{2}\tan^{-1}\left(\frac{2qv}{(1-v)+q^2(1+v)}\right)H(v). \end{aligned} \tag{5.59}$$

Thus, the polarization angle for the impulsive part of Ψ_0 remains constant even after the collision. The reason is that, similar to the last case, Ψ_0^{Im} does not interact with any of the other components of both gravitational and matter fields. However, for the Ψ_4 wave the situation is different. The interaction between the impulsive part of Ψ_4 and the impulsive shell of null dust Φ_{22}^{Im} is such as to make the polarization angle θ_4^{Im} change along the v-axis according to Eq. (5.59).

In a similar fashion, we find that for the shock part of Ψ_0 and Ψ_4 the polarization angles θ_0^{sh} and θ_4^{sh} are given, respectively, by

$$\begin{aligned} \theta_0^{\text{sh}} &= -\frac{1}{2}\tan^{-1}\left(\frac{I(v,Y)}{J(v,Y)}\right)H(v), \\ \theta_4^{\text{sh}} &= \frac{1}{2}\tan^{-1}\left(\frac{I(Y,v)}{J(Y,v)}\right)H(u)H(v), \end{aligned} \tag{5.60}$$

where the functions $I(x,y)$ and $J(x,y)$ are defined by

$$\begin{aligned} I(x,y) \equiv 2qy\{&(x^2-y^2)(x-y)(x+5y) + q^2[(4+6q^2+4q^4+q^6)x^4 \\ &+ 4(2-2q^4-q^6)x^3y + 2(4+14q^2+4q^4-3q^6)x^2y^2 \\ &+ 4(6-6q^4+q^6)xy^3 - (60-126q^2+60q^4-5q^6)y^4]\}, \end{aligned}$$

$$J(x,y) \equiv (x^2 - y^2)^2(x - y) + q^2\{(5 + 10q^2 + 10q^4 + 5q^6 + q^8)x^5$$
$$- (3 + 2q^2 - 2q^4 - 3q^6 - q^8)x^4y + 2(3 + 14q^2 + 14q^4$$
$$+ 3q^6 - q^8)x^3y^2 - 2(13 + 14q^2 - 14q^4 - 13q^6 + q^8)x^2y^3 \quad (5.61)$$
$$- (27 - 42q^2 - 42q^4 + 27q^6 - q^8)xy^4 + (45 - 210q^2$$
$$+ 210q^4 - 45q^6 + q^8)y^5\}.$$

Equation (5.60) shows that the plane gravitational shock waves of Ψ_0 and Ψ_4 change their polarization directions along each of their own paths. These changes are due to the presence of the Coulomb-like field Ψ_2 in Eq. (3.70), which is due to the nonlinear interaction between Ψ_0 and Ψ_4.

5.2. Collisions of Massless Scalar Waves

The gravitational interaction of colliding null dust clouds has been investigated by several authors (Chandrasekhar and Xanthopoulos, 1985b, 1985c, 1986b; Taub, 1988a, 1988b, 1990; Ferrari and Ibañez, 1989a, 1989b; Feinstein, MacCallum and Senovilla 1989; Tsoubelis and Wang, 1991). Many remarkable features of the interaction have been found. One of these arose initially from two solutions of the Einstein field equations constructed by Chandrasekhar and Xanthopoulos (1985b, 1986b). According to the above authors, one of the two solutions represents the collision of two null dust clouds (Chandrasekhar and Xanthopoulos, 1985b), each of which is accompanied by an impulsive gravitational plane wave. The product of the collision is a perfect fluid with the equation of state, p (pressure) $= \varepsilon$ (energy density). The other solution represents the collision of the same null dust clouds but with the product being a mixture of null dust moving in opposite directions (Chandrasekhar and Xanthopoulos, 1986b). The presence of two different fluids in the interaction region gave rise to the following issues. One concerns the uniqueness of the collision, and the other concerns the possibility of a gravitationally induced transformation (or phase transition) of massless particles into massive ones.

Taub (1988a, 1988b) showed that the different outcome of the above collision is due to the different assumptions on the Ricci tensor, which specify the nature of the interaction. In other words, in order to restore the uniqueness, one has to impose some conditions on the Ricci tensor. Feinstein, MacCallum and Senovilla (1989), on the other hand, argued that the ambiguities arise from an incomplete physical treatment of the problem, in the sense that only the gravitational field is considered in the

Chandrasekhar–Xanthopoulos approach. These ambiguities remain even after Taub's points are taken into account. They disappear only when the underlying physical equations of the matter fields are taken into considerations. They also argued that the transformation of colliding null dust into other forms of matter is unrealizable in classical general relativity.

Tsoubelis and Wang (1991), however, showed that as far as the uniqueness is concerned, the approach given by Taub is equivalent to the one given by Feinstein *et al.* Actually, if one follows the instructions given by Feinstein *et al.* that first specify the matter field and then solve both the Einstein field equations and the matter field equations, one will find that this is no more than to provide the conditions imposed by Taub.

The difference between the above two kinds of approaches essentially reflects the well-known "problem" of the right-hand side of the Einstein field equations (Hawking and Ellis, 1973) or in Synge's words, the problem between the realist and the agonist (Synge, 1965). The realist wants to connect the energy–stress tensor appearing in the right-hand side of the Einstein field equations (1.20) with the known matter fields. So, the starting point is first to specify the matter field, and then to solve the coupling Einstein and matter field equations. The agonist, on the other hand, wants to wrestle with the difficult mathematical problems arising out of the field equations. Then, the starting point is to construct the metric (i.e. to solve the left-hand side of the Einstein field equations) disregarding the physical constitution of matter. After the metric is given, the energy–stress tensor is read off from the Einstein field equations. However, such obtained energy-stress tensor does not always physically acceptable, as they commonly violate energy conditions (Hawking and Ellis, 1973). To be physically acceptable, the agonist has to impose some conditions on the metric or equivalently on the Ricci tensor.

In this section, by considering a class of solutions which represents the collision of massless scalar waves, we shall show that, besides the above mentioned problems, the fluid picture has also difficulties in interpreting some results of the above kind of collisions. But, these difficulties can be overcome if we consider the collision as the one of two massless scalar waves.

Before proceeding, we first present a theorem, which is very useful in constructing the above kind of collisions from the known solutions of the Einstein vacuum equations.

First we notice that the Einstein field equations when coupled with a massless scalar field are given by Eqs. (3.8a)–(3.8e) with the non-vanishing Ricci scalars given by Eqs. (3.67) and (3.69). Following Wainwright, Ince and Marshman (1979), Tabensky and Taub (1973) and Taub (1988b), we find that solutions of a massless scalar field can be obtained from vacuum ones by simply setting

$$(M, U, V, W) = (M_v + \Omega, U_v, V_v, W_v), \tag{5.62}$$

where M_v, U_v, V_v and W_v are solutions of the Einstein vacuum equations, and the function Ω satisfies

$$\Omega_{,u} = \frac{\phi_{,u}}{U_{,u}}, \tag{5.63a}$$

$$\Omega_{,v} = \frac{\phi_{,v}}{U_{,v}}, \tag{5.63b}$$

$$\Omega_{,uv} = \phi_{,u}\phi_{,v}, \tag{5.63c}$$

where the scalar field ϕ satisfies the massless scalar field equation (3.80). If we consider, on the other hand, the metric

$$ds^2 = 2e^{-M^{(0)}} du dv - e^{-U}[e^{V^{(0)}}(dx^2)^2 + e^{-V^{(0)}}(dx^3)^2], \tag{5.64}$$

we find that the corresponding Einstein vacuum equations are given by [see Eqs. (3.8a)–(3.8e)],

$$2U_{,vv} - U_{,v}^2 + 2U_{,v}M_{,v}^{(0)} = V_{,v^2}^{(0)}, \tag{5.65a}$$

$$2U_{,uu} - U_{,u^2} + 2U_{,u}M_{,u}^{(0)} = V_{,u^2}^{(0)}, \tag{5.65b}$$

$$2M_{,uv}^{(0)} = U_{,u}U_{,v}, \tag{5.65c}$$

and

$$U_{,uv} - U_{,u}U_{,v} = 0, \tag{5.66}$$

$$2V_{,uv}^{(0)} - U_{,u}V_{,v}^{(0)} - U_{,v}V_{,u}^{(0)} = 0. \tag{5.67}$$

Introducing the function $\Omega^{(0)}$ by

$$M(0) = N^{(0)} + \Omega^{(0)}, \tag{5.68}$$

where

$$N^{(0)} \equiv \frac{3}{2}U - \ln|2U_{,u}U_{,v}|, \tag{5.69}$$

we find that Eqs. (5.65a)–(5.65c) can be written as

$$\Omega_{,u}^{(0)} = \frac{V_{,u}^{(0)}}{2U_{,u}}, \quad \Omega_{,v}^{(0)} = \frac{V_{,v}^{(0)}}{2U_{,v}},$$

$$\Omega_{,uv}^{(0)} = \frac{1}{2}V_{,u}^{(0)}V_{,v}^{(0)}. \tag{5.70}$$

Comparison of Eqs. (3.80) and (5.63a)–(5.63c) with Eqs. (5.67) and (5.70) leads to the following theorem.

Theorem (Tsoubelis and Wang, 1991). *Let $(M^{(0)}, U, V^{(0)})$ be a solution of the Einstein vacuum equations corresponding to the metric (5.64), Then*

$$(M, U, V, W) = (M_v + \lambda^2(M^{(0)} - N^{(0)}) + C_0, U, V_v, W_v), \qquad (5.71)$$

with $N^{(0)}$ given by Eq. (5.69) is a solution of the Einstein equations coupled with a massless scalar field $\phi = \lambda V^{(0)}/\sqrt{2}$, where λ and C_0 are constant.

Note that when $V_v = W_v = 0$, the Einstein vacuum equations give the following solution for the function M_v:

$$M_v = N^{(0)} + C_1, \qquad (5.72)$$

where C_1 is another arbitrary constant.

As an application of the above theorem, we use the vacuum solutions given by Eqs. (4.23) and (4.24) as the solutions of $(M^{(0)}, U, V^{(0)})$, and for the sake of convenience we set the functions V_v and W_v to be zero

$$V_v = W_v = 0. \qquad (5.73)$$

Then, the function M_v has the solution given by Eq. (5.72). By choosing the parameter λ introduced in Eq. (5.71) to be $\sqrt{2}$, we finally obtain the following solutions:

$$\phi = V^{(0)} \equiv a \ln[(1 - \eta^2)(1 - \mu^2)] + \delta_1 \ln \left(\frac{1 - \eta}{1 + \eta} \right)$$

$$+ \delta_2 \ln \left(\frac{1 - \mu}{1 + \mu} \right), \qquad (5.74)$$

and

$$e^{-U} = 1 - u^{2n} - v^{2n},$$

$$e^{-M} = (1 - \eta)^{b_1}(1 - \mu)^{b_2}(1 + \eta)^{c_1}(1 + \mu)^{c_2} X^{-2\delta_+^2} Y^{-2\delta_-^2}, \qquad (5.75)$$

$$V = W = 0,$$

but now b_A and c_A are defined by

$$b_A = 2(a + \delta_A)^2 - \frac{1}{4}, \quad c_A = 2(a - \delta_A)^2 - \frac{1}{4}, \ (A = 1, 2), \qquad (5.76)$$

and the parameters n and m are related to $\delta_\pm$ via the relations,

$$n = \frac{1}{2(1 - \delta_+^2)}, \quad m = \frac{1}{2(1 - \delta_-^2)}. \qquad (5.77)$$

As usual, taking the above solutions as valid only in the interaction region (Region IV), we can extend them to the pre-collision regions by means of the Khan–Penrose substitutions (3.11). Specifically, the pair (u, v) is replaced by the one $(uH(u), vH(v))$. However, the results obtained in the last chapter and the ones in Section 5.1 show that the extended solutions are physically acceptable only in the case where

$$n = \frac{1}{2}, \text{ or } \geq 1, \quad \text{and} \quad m = \frac{1}{2}, \text{ or } \geq 1. \tag{5.78}$$

It can be shown that such extended function ϕ still satisfies the massless scalar field equation (3.80).

As shown in Section 5.1, the above extension leads an impulsive shell of null dust to appear on the hypersurface $u = 0(v = 0)$ if $n = 1/2$ $(m = 1/2)$.

We note that the above three-parameter class of solutions is a member of the Ferrari–Ibañez four-parameter class of solutions (Ferrari and Ibañez, 1989a), and was described as "stiff matter" solutions. However, the following considerations show that such a term does not justice to the physically rich class of models given by Eqs. (5.74) and (5.75).

To begin with, we first show that a massless scalar field is not always energetically equivalent to a perfect fluid with the equation of state $p = \varepsilon$. Writing the energy–stress tensor given by Eq. (1.92) in the form

$$T_{\mu\nu} = \phi_{,\mu}\phi_{,\nu} - \varepsilon g_{\mu\nu}, \tag{5.79}$$

where

$$\varepsilon \equiv \frac{1}{2}\phi_{;\mu}\phi^{;\mu}, \tag{5.80}$$

we find that the following possibilities in general arise (Tabensky and Taub, 1973).

Case (i): $\varepsilon > 0$. In this case, the vector $\phi_{,\mu}$ is time-like. If we introduce the unit vector T_μ by

$$T_\mu \equiv \frac{\phi_{,\mu}}{(2\varepsilon)^{1/2}}, \tag{5.81}$$

we find that the metric and the energy–stress tensor can be written in the form

$$g_{\mu\nu} = T_\mu T_\nu - Z_\mu Z_\nu - X_\mu X_\nu - Y_\mu Y_\nu, \tag{5.82}$$

$$T_{\mu\nu} = \varepsilon(T_\mu T_\nu + Z_\mu Z_\nu + X_\mu X_\nu + Y_\mu Y_\nu), \tag{5.83}$$

where X_μ, Y_μ and Z_μ are three orthogonal unit space-like vectors, which, together with T_μ, consist of an orthogonal tetrad. Then, Eq. (5.83) implies that the scalar field ϕ is energetically equivalent to a perfect fluid with four-velocity proportional to $\phi_{,\mu}$ and the equation of state, $p = \varepsilon$, which is often referred to as the "still" fluid.

Case (ii): $\varepsilon < 0$. In this case, $\phi_{,\mu}$ defines a space-like unit vector, say Z_μ, via the relation

$$Z_\mu = \frac{\phi_{,\mu}}{(-2\varepsilon)^{1/2}}. \tag{5.84}$$

With similar considerations, we find that the metric in the present case is given by Eq. (5.82), while the energy–stress tensor is given by

$$T_{\mu\nu} = |\varepsilon|(T_\mu T_\nu + Z_\mu Z_\nu - X_\mu X_\nu - Y_\mu Y_\nu). \tag{5.85}$$

Therefore, in this case the scalar field ϕ is energetically equivalent to an anisotropic fluid with vanishing heat flow.

Case (iii): $\varepsilon = 0$: In this case, $\phi_{,\mu}$ is a null vector and $T_{\mu\nu}$ takes the form that we have been calling a null dust cloud propagating in the direction defined by $\phi^{,\mu}$.

The significance of the above observations will be made clear as we are going to reveal the fact that in the solutions under considerations all the above possibilities arise in the interaction region (Region IV) [see Fig. 5.1].

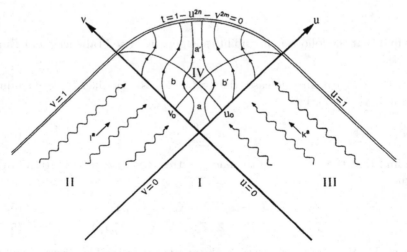

Fig. 5.1.   The projection of the colliding null dust spacetime onto the (u, v)-plane.

In particular, in this region we have

$$\varepsilon = e^M R_{uv}$$

$$= \frac{4nme^M \tan \zeta \tan \xi}{t^2 (\sin \zeta)^{1/n} (\sin \xi)^{1/m}} [a \sin 2\zeta + \delta_+ \cos \zeta \cos \xi + \delta_- \sin \zeta \sin \xi]$$

$$\times [a \sin 2\xi + \delta_- \cos \zeta \cos \xi + \delta_+ \sin \zeta \sin \xi], \tag{5.86}$$

where

$$\sin \zeta \equiv u^n, \quad \sin \xi \equiv v^m. \tag{5.87}$$

Equation (5.86) shows that to study the above solutions as a whole is very complicated, so in the following we consider only some representative cases.

Case A: $\delta_A = 0$ ($n = m = 1/2$). In this case, the scalar field ϕ is given by

$$\phi = a \ln[(1 - \eta^2)(1 - \mu^2)] = 2a \ln[1 - uH(u) - vH(v)], \tag{5.88}$$

while the combination of Eqs. (1.20), (1.92) and (5.88) yields,

$$R_{uu} = \frac{4a^2}{[1 - u - vH(v)]^2} H(u) + \frac{1}{1 - vH(v)} \delta(u), \tag{5.89a}$$

$$R_{vv} = \frac{4a^2}{[1 - uH(u) - v]^2} H(v) + \frac{1}{1 - uH(u)} \delta(v). \tag{5.89b}$$

After the collision, the component R_{uv} appears, which makes the scalar field energetically equivalent to a perfect fluid with the energy density and pressure given by

$$p = \varepsilon = \frac{4a^2}{(1 - u - v)^{(8a^2+3)/2}} = \frac{4a^2}{t^{(8a^2+3)/2}}. \tag{5.90}$$

The last equation shows explicitly that this class of solutions develops a spacetime curvature singularity along the hypersurface $t = 0$. In fact, the singularity happens in all the colliding wave models considered in this section.

Case B: $a = \delta_1 = 0$. In this case, we have

$$\phi = \delta \ln\left(\frac{1-\mu}{1+\mu}\right), \quad \delta \equiv \delta_2, \tag{5.91}$$

and

$$R_{uu}(v < 0) = 4n^2\delta^2[1 - u^{2n}]^{-2}u^{2(n-1)}H(u), \tag{5.92a}$$

$$R_{vv}(u < 0) = 4m^2\delta^2[1 - v^{2m}]^{-2}v^{2(m-1)}H(v), \tag{5.92b}$$

$$R_{uv}(u, v) = -\frac{4n^2\delta^2(\sin\zeta\sin\xi)^{2\delta^2-1}\cos^2(\zeta+\xi)}{t^2\cos\zeta\cos\xi}H(u)H(v), \tag{5.92c}$$

where

$$n = m = \frac{1}{2(1-\delta^2)} \geq 1. \tag{5.93}$$

From Eqs. (5.92a)–(5.92c) we find that in this case the solutions represent the collision of two null dust clouds only. Unlike the last case, the collision and interaction of the two null dust clouds are such that in the interaction region (Region IV), the function ε defined by Eq. (5.80) is no longer positive as follows from Eq. (5.92c). Therefore, in the present case the scalar field ϕ in Region IV is energetically equivalent to an anisotropic fluid with vanishing heat-flow.

Case C: $\delta_1^2 \equiv \delta^2 = 1/2, \delta_2 = 0$. In this case, we have $n = m = 1$. Restricting ourselves again to the physically relevant quantities, we find that

$$\phi = 2a\ln[1 - u^2H(u) - v^2H(v)] + \delta\ln\left(\frac{1-\eta}{1+\eta}\right), \tag{5.94}$$

and

$$R_{\mu\nu}(u < 0) = 4\left(\frac{\delta + 2av}{1 - v^2}\right)^2 H(v)\hat{n}_\mu\hat{n}_\nu, \tag{5.95a}$$

$$R_{\mu\nu}(v < 0) = 4\left(\frac{\delta + 2au}{1 - u^2}\right)^2 H(u)\hat{l}_\mu\hat{l}_\nu, \tag{5.95b}$$

$$R_{\mu\nu}(u, v) = \frac{4H(u)H(v)}{t^2\cos\zeta\cos\xi}[a\sin 2\zeta + \delta\cos(\zeta - \xi)]$$
$$[a\sin 2\xi + \delta\cos(\zeta - \xi)]. \tag{5.95c}$$

As shown by Eqs. (5.95a) and (5.95b), the present class of solutions also represents the collision of two pulses of scalar waves which are free of impulsive components. On the other hand, Eq. (5.95c) shows that the superposition of the incoming waves after collision is such that one cannot regard the interaction region as consisting of a fluid of a certain type. In particular, assuming that $\delta a < 0$ with $|\delta/2a| < 1$, from Eqs. (5.95a)–(5.95c) we find that $R_{uv} = 0$ along the following two curves in Region IV. The first is defined by the relation

$$\sin 2\zeta = -\frac{\delta}{a}\cos(\zeta - \xi), \qquad (5.96)$$

and has its origin at the point $(u_0, 0)$ with $u_0 = |\delta/2a|$. Similarly, the second curve is defined by

$$\sin 2\xi = -\frac{\delta}{a}\cos(\zeta - \xi), \qquad (5.97)$$

and starts at $(0, v_0)$ where $v_0 = u_0$. Across these curves the sign of R_{uv} changes. Therefore, Region IV is split into subregions occupied by a fluid whose type alternates among the cases (i) and (ii) considered after Eq. (5.80), as illustrated in Fig. 5.1.

Case D: $a = 0, \delta_A = \delta \neq 0$. In this case, we find that

$$\phi = \delta \ln \frac{(1-\eta)(1-\mu)}{(1+\eta)(1+\mu)}, \qquad (5.98)$$

and

$$R_{uu} = \frac{16n^2\delta^2}{t^2}Y^2 u^{2(n-1)}H(u), \qquad (5.99a)$$

$$R_{vv} = \frac{4\delta^2}{t^2Y^2}u^{2n}H(u)H(v) + \frac{1}{1-u^{2n}H(u)}\delta(v), \qquad (5.99b)$$

$$R_{uv} = \frac{8n\delta^2}{t^2}u^{2n-1}H(u)H(v), \qquad (5.99c)$$

which show that in the present case the solutions represent the collision of a null dust cloud with an impulsive shell of null dust. After the collision, a current moving along the $v =$ constant lines toward the right-hand side is developed, which together with the R_{uv} component given by Eq. (5.99c) again makes the scalar field in the interaction region energetically equivalent to a perfect fluid with the equation of state, $p = \varepsilon$.

However, further considerations (Tsoubelis and Wang, 1991) showed that the origin of the right-hand moving current is difficult to understand if one sticks to the fluid picture, since before the collision there is no incoming current moving toward the right-hand side as follows from Eq. (5.99c). Instead, if we consider the question in terms of the scalar field, from the Klein–Gordon equation (3.80) we can see that this is because of the backscattering. Actually, in terms of the latter, the present model represents the collision of a scalar plane wave incident from the right-hand side, which collides at $(u, v) = (0, 0)$ with an impulsive wave (shell of null dust). The focusing that results from this encounter is equivalent to the scalar wave pulse's entering a spacetime region where the metric depends on both u and v. But, then the Klein–Gordon equation (3.80) implies that right moving waves are generated, i.e. backscattering occurs. This explains the appearance of the right moving current in Region IV in terms of the null current picture.

5.3. Collision of Neutrino and Electromagnetic Waves

To further discuss the issues of the uniqueness of the initial value problem of colliding null dust clouds and the gravitational "phase transitions" induced from the collision of matter fields, in this section we consider the spacetime described by the metric

$$ds^2 = 2e^{-M}dudv - e^{-U}[(dx^2)^2 + (dx^3)^2], \tag{5.100}$$

subject to the assumptions

$$U_{,u}U_{,v} \neq 0, \tag{5.101}$$

$$R_{22} = R_{33} = 0, \tag{5.102}$$

$$R_{uv} = R_{vu} = 0, \tag{5.103}$$

in order to make the analysis as simple as possible, and meantime the relevant physics still remains. Then, the combination of Eqs. (3.8a)–(3.8e), (5.100), (5.103) and the Einstein field equations (1.20) yields

$$(e^{-U})_{,uv} = -e^{-U}(U_{,uv} - U_{,u}U_{,v}) = 0, \tag{5.104}$$

which, together with Eq. (5.101), gives the following solution:

$$e^{-U} = \alpha(u) + \beta(v), \tag{5.105}$$

where $\alpha(u)$ and $\beta(v)$ are arbitrary functions of their indicated arguments. On the other hand, from Eqs. (3.8a)–(3.8e), (4.20a) and (4.20b) we find that corresponding to the metric (5.100) the condition (5.103) now reads

$$2M_{,uv} + U_{,uv} = 0. \tag{5.106}$$

For the function U given by Eq. (5.105), we find that Eq. (5.103) has the general solution

$$M = -\frac{1}{2}U - A(u) - B(v), \tag{5.107}$$

where A, B are other two arbitrary functions of their indicated arguments.

Corresponding to the metric (5.100) with the function M and U given by Eqs. (5.107) and (5.105), it can be shown that the Ricci tensor $R_{\mu\nu}$ is given by

$$R_{\mu\nu} = \frac{\alpha' A' - \alpha''}{\alpha + \beta}\hat{l}_\mu \hat{l}_\nu + \frac{\beta' B' - \beta''}{\alpha + \beta}\hat{n}_\mu \hat{n}_\nu, \tag{5.108}$$

where the prime denotes the ordinary differentiation with respect to the indicated argument. Thus, the corresponding energy–stress tensor can be written in the form

$$T_{\mu\nu} = T^L_{\mu\nu} + T^R_{\mu\nu}, \tag{5.109}$$

where

$$T^L_{\mu\nu} = \varepsilon^L \hat{l}_\mu \hat{l}_v \equiv \frac{\alpha' A' - \alpha''}{\alpha + \beta}\hat{l}_\mu \hat{l}_v, \tag{5.110a}$$

$$T^R_{\mu\nu} = \varepsilon^R \hat{n}_\mu \hat{n}_v \equiv \frac{\beta' B' - \beta''}{\alpha + \beta}\hat{n}_\mu \hat{n}_v. \tag{5.110b}$$

Equations (5.109) and (5.110a)–(5.110b), together with Eq. (1.21), yield

$$T^{L;\nu}_{\mu\nu} = 0 = T^{R;\nu}_{\mu\nu}. \tag{5.111}$$

It follows that, assuming both ε^L and ε^R be positive, the source given by Eqs. (5.109) and (5.110a)–(5.110b) consists of a pair of oppositely moving null dust clouds, one of which has the conserved energy density ε^L, while the other has the conserved energy density ε^R.

In order to consider the solutions representing the collision of two null dust clouds, let us assume the functions A and B introduced in Eq. (5.107) such that

$$A(0) = B(0) = 0. \tag{5.112}$$

We also assume that α and β, which specify the function U via Eq. (5.105), are chosen such that

$$\alpha(0) = \beta(0) = \frac{1}{2}. \tag{5.113}$$

Then, we can see that the conditions (3.22) are now satisfied, and the Khan–Penrose substitutions (3.11) can be used. Replacing the pair (u, v) by the one $(uH(u), vH(v))$ in Eqs. (5.105) and (5.107), we find that the extended solutions are given by

$$e^{-M} = \frac{e^{A(u)H(u)+B(v)H(v)}}{[1 - f(u)H(u) - g(v)H(v)]^{1/2}}, \tag{5.114}$$
$$e^{-U} = 1 - f(u)H(u) - g(v)H(v),$$

where

$$f(u) = \alpha(0) - \alpha(u), \quad g(v) = \beta(0) - \beta(v). \tag{5.115}$$

Consequently, the corresponding energy–stress tensor is now given by

$$T_{\mu\nu} = e^{U}[(f'' - f'A')H(u)\hat{l}_{\mu}\hat{l}_{\nu} + (g'' - g'B')H(v)\hat{n}_{\mu}\hat{n}_{\nu}$$
$$+ f'\delta(u)\hat{l}_{\mu}\hat{l}_{\nu} + g'\delta(v)\hat{n}_{\mu}\hat{n}_{\nu}]. \tag{5.116}$$

The only non-vanishing Weyl scalar, on the other hand, is given by

$$\Psi_2 = \frac{1}{2}(M_{,uv} - U_{,uv}) = -\frac{1}{4}e^{2U}f'g'H(u)H(v), \tag{5.117}$$

which means that all of the models under consideration are conformally flat, except for the ones in which $f'g' \neq 0$.

Now let us turn to prove that the two non-impulsive terms appearing in Eq. (5.116) for $T_{\mu\nu}$ can be interpreted as representing an electromagnetic plane wave and a neutrino plane wave, respectively. To this purpose, let us consider the antisymmetric tensor $F_{\mu\nu}$, which is defined by

$$F_{\mu\nu} = 2\Phi_0\hat{n}_{[\mu}m_{\nu]} + 2\overline{\Phi}_0\hat{n}_{[\mu}\overline{m}_{\nu]}, \tag{5.118}$$

where $\Phi_0 = \Phi_0(u, v)$, and m_μ is given by

$$m_\mu = \frac{1}{\sqrt{2}}e^{-U/2}(\delta_\mu^2 + i\delta_\mu^3). \tag{5.119}$$

Provided that it satisfies the Maxwell equation (1.96), $F_{\mu\nu}$ represents a null electromagnetic field with the energy–stress tensor $T_{\mu\nu}$ given by

$$T_{\mu\nu}^e = F_{\mu\lambda}F^\lambda{}_\nu - \frac{1}{4}g_{\mu\nu}F_{\rho\lambda}F^{\lambda\rho} = 2\Phi_0\overline{\Phi}_0\hat{n}_\mu\hat{n}_\nu. \tag{5.120}$$

Comparing Eq. (5.116) with Eq. (5.120), we conclude that the second term in the right-hand side of Eq. (5.116) can be attributed to an electromagnetic wave, if the function Φ_0 obtained from

$$\Phi_0\overline{\Phi}_0 = \frac{1}{2}e^U(g'' - g'B')H(v),$$ (5.121)

satisfies the Maxwell equation (1.96) through Eq. (5.118). In fact, choosing Φ_0 such that

$$\Phi_0 = \left[\frac{1}{2}e^U(g'' - g'B')\right]^{1/2}H(v),$$ (5.122)

we find that both Eq. (5.120) and the Maxwell field equation (1.96) are satisfied in all the four regions (Regions I–IV) as well as on the separation null hypersurfaces Σ_u and Σ_v. Substituting Eq. (5.122) into Eq. (5.118), we find

$$F_{\mu\lambda} = 2\sqrt{g'' - g'B'}H(v)\delta^1_{[\mu}\delta^2_{\nu]}.$$ (5.123)

On the other hand, if we set $\Psi = 0$ in Eq. (1.102), we can show that the only non-vanishing Ricci scalar for a neutrino field is Φ_{22} [see Eq. (1.105)]. Introducing the function $\Phi^{(0)}$ by

$$\Phi = A^{1/2}\Phi^{(0)},$$ (5.124)

where A is the function introduced in the null tetrad (2.23) related to the function M via Eq. (2.25), we find that the Weyl neutrino equations (1.104a) and (1.104b) reduce to a single equation,

$$\Phi^{(0)}_{,v} = \frac{1}{2}U_{,v}\Phi^{(0)},$$ (5.125)

and that the corresponding energy–stress tensor takes the form,

$$T^n_{\mu\nu} = i2[\Phi^{(0)}\overline{\Phi^{(0)}}_{,u} - \overline{\Phi^{(0)}}\Phi^{(0)}_{,u}]\hat{l}_\mu\hat{l}_\nu.$$ (5.126)

The solution of Eq. (5.125) can be written in the form

$$\Phi^{(0)} = e^{U/2}\rho(u)e^{i\theta(u)}.$$ (5.127)

A particular solution of the functions ρ and θ is given by

$$\rho(u) = \frac{1}{2}(f'' - f'A')^{1/2}H(u), \quad \theta(u) = u.$$ (5.128)

The combination of Eqs. (5.127), (5.128) and (5.126) yields,

$$T^n_{\mu\nu} = e^U(f'' - f'A')H(u)\hat{l}_\mu\hat{l}_\nu.$$ (5.129)

This expression now can be compared with the first term on the right-hand side of Eq. (5.116), from which we conclude that this term can be considered as representing a neutrino plane wave, which, according to Eqs. (1.102), (5.124), (5.127) and (5.128), is given by

$$\phi_A = \frac{1}{2}[A(f'' - f'A')e^U]^{1/2}H(u)e^{iu}O_A. \tag{5.130}$$

Having completed the proof that the above extended solutions can be interpreted as representing the collision of a pulse of electromagnetic radiation incident from the left in Fig. 5.1 with a pulse of neutrino radiation incident from the right, let us turn to consider some specific examples.

We first choose the functions f and g to be of the forms,

$$f(u) = u^{2n}, \quad g(u) = v^{2m}, \tag{5.131}$$

where n and m satisfy conditions (5.78). Then, we choose the functions A and B to be of the forms,

$$\begin{aligned} A &= p\ln(1 - u^n) + q\ln(1 + u^n), \\ B &= r\ln(1 - v^m) + s\ln(1 + v^m), \end{aligned} \tag{5.132}$$

with the tetrad of constants (p, q, r, s) being chosen so that

$$\begin{aligned} p = q \geq 0, \quad \text{when } n = \frac{1}{2}, \\ r = s \geq 0, \quad \text{when } m = \frac{1}{2}. \end{aligned} \tag{5.133}$$

Otherwise, these four constants have to be chosen so that the following conditions always hold for any $x \in (0, 1)$:

$$\begin{aligned} F(x) &= [n(p + q) - (2n - 1)]x^{2n} + n(p - q)x^n + 2n - 1 \geq 0, \\ G(x) &= [m(r + s) - (2m - 1)]x^{2m} + m(r - s)x^m + 2m - 1 \geq 0. \end{aligned} \tag{5.134}$$

Then, Eqs. (5.132)–(5.134) define a six-parameter family of solutions corresponding to sources described by the energy–momentum tensor,

$$T_{\mu\nu} = T^e_{\mu\nu} + T^n_{\mu\nu} + \frac{2mv^{2m-1}\delta(v)}{1 - u^{2n}H(u)}\hat{n}_\mu\hat{n}_\nu + \frac{2nu^{2n-1}\delta(u)}{1 - v^{2m}H(v)}\hat{l}_\mu\hat{l}_\nu, \tag{5.135}$$

where

$$T^e_{\mu\nu} = \frac{2mv^{2m-2}G(v)H(v)}{(1 - v^{2m})[1 - u^{2n}H(u) - v^{2m}]}\hat{n}_\mu\hat{n}_\nu, \tag{5.136}$$

$$T^n_{\mu\nu} = \frac{2nu^{2n-2}F(u)H(u)}{(1 - u^{2n})[1 - u^{2n} - v^{2m}H(v)]}\hat{l}_\mu\hat{l}_\nu, \tag{5.137}$$

with $G(v)$ and $F(u)$ being given by Eq. (5.134). The non-impulsive parts of the above sources can be attributed to a pulse of electromagnetic wave and a pulse of neutrino wave given, respectively, with the field strengths,

$$F_{\mu\nu} = 2v^{m-1} \left[\frac{2mG(v)}{1 - v^{2m}} \right]^{1/2} H(v)\overline{n}_{[\mu}\delta^2_{\nu]}, \qquad (5.138)$$

$$\phi_A = \frac{u^{n-1}}{2} \left[\frac{2nAF(u)e^U}{1 - u^{2n}} \right]^{1/2} H(u)e^{iu}O_A. \qquad (5.139)$$

The above class of solutions covers a large variety of physically interesting cases, which are obtained by choosing the corresponding parameters appropriately. When the tetrad (p, q, r, s) vanishes and $n = m = 1$, for example, we recover Griffiths' solution (Griffiths, 1976b), which represents the collision of a pair of constant profile shock waves made of photons and neutrinos, respectively. When $n = 2$ and $m = 1$, the corresponding solutions represent the collision of an electromagnetic shock wave and a neutrino wave with smooth wavefront, etc.

Without going to more detailed analyses for each of specific models, we turn to the issues, regarding to the ambiguity of the outcome of collisions of matter fields, raised in Section 5.2. Choosing the parameters p, q, r, s, n and m as

$$p = (2a + \delta_+)^2, \quad q = (2a - \delta_+)^2, \quad r = (2a + \delta_-)^2, \quad s = (2a - \delta_-)^2,$$
$$n = [2(1 - \delta_+)^2]^{-1}, \quad m = [2(1 - \delta_-)^2]^{-1}, \qquad (5.140)$$

where a and δ_A will be chosen as the same constants that specify the family of scalar wave solutions analyzed in the last section, we obtain a three-parameter family of the solutions given by Eqs. (5.131)–(5.134). It is then easy to verify that in the pre-collision regions (Regions I–III), the metric resulting from the choice of Eq. (5.140) is identical to the one corresponding to the scalar wave models studied in Section 5.2, even though the solutions differ from each other in the interaction region (Region IV). Thus, Taub's results about the ambiguity from planar colliding null dust clouds with impulsive waves now extend to solutions that are free from such impulsive components. That is, given the metric (solutions of the Einstein field equations) in Regions I–III, one would not be able to predicate what is the outcome of the collision. The colliding process becomes determinative, however, once the energy–stress tensor or equivalently, the type of interaction is pre-assigned in Region IV. In the present case, the resolution of the ambiguity was resolved by constructing the corresponding energy–stress distributions from the well-defined matter fields, neutrino and electromagnetic.

5.4. Collisions of Two Electromagnetic Plane Waves

The studies of collisions of two electromagnetic plane waves can be traced back to the seminal work of Bell and Szekeres (1974), in which they presented a solution that represents the collision of two pure plane electromagnetic shock waves, and the metric in the interaction region takes the form,

$$
ds_{\mathrm{IV}}^2 = 2dudv - [1 - \sin^2(au) - \sin^2(bv)] \left[\frac{\cos(au - bv)}{\cos(au + bv)} (dx^2)^2 \right.
$$

$$
\left. + \frac{\cos(au + bv)}{\cos(au - bv)} (dx^3)^2 \right],
\tag{5.141}
$$

where a and b are two real constants. It is interesting to note that in the current case the focusing surface, $1 - \sin^2(au) - \sin^2(bv) = 0$ (or $au + bv = \pi/2$), is not singular, but a Cauchy horizon (Bell and Szekeres, 1974). Clarke and Hayward (1989) showed that extensions beyond this surface are not unique, and similar to the vacuum case (Yurtsever, 1987), it was argued that such a horizon is also not stable (Konkowski and Helliwell, 1991; Gürses and Halilsoy, 1982; Gurtug and Halilsoy, 2000). Another interesting aspect of the solution is that metric (5.141) is conformally flat,

$$
ds_{\mathrm{IV}}^2 = \frac{q^2}{r^2} [dt^2 - dr^2 - r^2(d\theta^2 + \sin^2\theta d\phi^2)],
\tag{5.142}
$$

where $2ab = q^{-2}$, and the mapping between the two sets of coordinates is given explicitly by (Griffiths, 1991),

$$
t + r = \coth\left[\frac{1}{2}\mathrm{sech}^{-1}\cos(au + bv) - \frac{y}{2q} \right],
$$

$$
t - r = -\tanh\left[\frac{1}{2}\mathrm{sech}^{-1}\cos(au + bv) + \frac{y}{2q} \right],
\tag{5.143}
$$

$$
\theta = au - bv + \frac{\pi}{2}, \quad \phi = \frac{1}{2}x.
$$

Using the Khan–Penrose substitutions, one can easily find the metric in Regions I–III,

$$
ds^2 = \begin{cases} 2dudv - \cos^2(au)d\Sigma_0^2, & \text{Region III}, \\ 2dudv - \cos^2(bv)d\Sigma_0^2, & \text{Region II}, \\ 2dudv - d\Sigma_0^2, & \text{Region I}, \end{cases}
\tag{5.144}
$$

where $d\Sigma_0^2 \equiv (dx^2)^2 + (dx^3)^2$. Clearly, Region I is flat, while in each of the two regions (Regions II and III) an electromagnetic shock wave is present,

$$\Phi_0 = bH(v), \quad \Phi_2 = aH(u), \tag{5.145}$$

while all the Weyl scalars Ψ_A $(A = 0, 2, 4)$ vanish in these regions as well as across their boundaries $u = 0$ and $v = 0$. Thus, the Bell–Szekeres solution represents the collision of two pure electromagnetic plane waves. However, as in the mixed cases presented in the last section, two impulsive gravitational waves are created due to the collision, one moves along the hypersurface $u = 0$ and the other along the one $v = 0$ (Bell and Szekeres, 1974),

$$\Psi_0 = -b\tan(au)H(u)\delta(v), \quad \Psi_4 = -a\tan(bv)H(v)\delta(u). \tag{5.146}$$

In the general case, after Chandrasekhar and Xanthopoulos (1985a) laid down the foundation of the problem, various solutions were obtained subsequently (Griffiths, 1985, 1990; Halilsoy, 1988a, 1988b, 1989a, 1989b; García-Díaz, 1988, 1989; Papacostas and Xanthopoulos, 1989; Li and Ernst, 1989; Bretón *et al.*, 1998; Hogan, Barrabés and Bressange, 1998; Hogan and Walsh, 2003). In particular, Chandrasekhar and Xanthopoulos (1985a) showed explicitly how one can generalize the Khan–Penrose and Nutku–Halii solutions to include electromagnetic plane shock waves, in which a spacetime curvature singularity is finally formed on the focusing surface. They also provided the generalization of the Bell–Szekeres solution to the case with two non-collinearly polarized electromagnetic plane shock waves (Chandrasekhar and Xanthopoulos, 1987b), and showed that a Cauchy horizon is finally formed on the focusing surface.

5.5. Other Colliding Plane Waves

In the last three decades or so, colliding plane waves have been studied extensively in various contents and theories of gravity (Griffiths, 1991; Stephani *et al.*, 2009; Griffiths and Podolsky, 2009). These include the dilaton gravity (Schwarz, 1997; Gürses and Sermutlu, 1995; Bretón, Matos and García, 1996; Gurtug, Halilsoy and Sakalli, 2003; Halilsoy and Sakalli, 2003), and high-dimensional spacetimes in the framework of string theory (Mizoguchi, 1996; Feistein, 2002; Gutperle and Pioline, 2003; Chen, 2004; Chen *et al.*, 2004; Chen and Zhang, 2005; Tziolas, Wang and Wu, 2009), as well as in the brane-worlds (de Leon, 2004; Tziolas and Wang, 2008). Many interesting results were obtained, including the possibility of disappearing of spacetime curvature singularities when left to higher dimensional spacetimes (Tziolas, Wang and Wu, 2009).

In the framework of string/M-theory, the AdS/CFT correspondence was investigated in the spacetimes of colliding gravitational shock waves (Rosenbaum *et al.*, 1986; Grumiller and Romatschke, 2008; Álvarez-Gaumé, *et al.*, 2009; Dueñas-Vidal and Vázquez-Mozo, 2010, 2012; Chesler and Yaffe, 2011; Chesler, 2015).

In addition, the collision of plane-fronted gravitational waves was also studied in curved backgrounds rather than the Minkowski. These include an expanding universe (Centrella and Matzner, 1982; Alekseev, 2016) and asymptotically flat spacetimes (Pretorius and East, 2018). Among other properties, it was found that a black hole can be formed due to the mutual focus of these plane-fronted waves (Pretorius and East, 2018), despite the fact that the total energy of the incoming waves now can be finite, in contrast to the uniform plane-fronted case, in which the total energy of the incoming waves is always infinitely large.

Chapter 6

Isometries to Interiors of Black Holes

In the past three decades or so, the collisions of gravitational plane waves coupled with various matter fields have been extensively studied, and very important insight about the nonlinearity of the Einstein field equations has been obtained.

One of the remarkable features is that the internal spacetimes of all the known black holes (Stephani *et al.*, 2009) are isometric to the interacting regions of two colliding plane gravitational waves with or without the presence of matter fields. The fundamental reason is that in both cases the Einstein field equations can be written in terms of the Ernst equation (3.44) (Chandrasekhar and Ferrari, 1984).[1] So, in this chapter we shall present a brief review on these fascinating phenomena.

6.1. The Schwarzschild Black Hole

The internal spacetime of the Schwarzschild black hole is locally isometric to the interacting region of two pure plane gravitational impulsive + shock waves (Ferrari and Ibañez, 1987a, 1987b; Yurtsever, 1988a, 1988b; Tsoubelis and Wang, 1989; Griffiths, 1991). In fact, setting

$$a = -\frac{1}{2}, \quad \delta_1 = 1, \quad \delta_2 = 0, \tag{6.1}$$

---

[1]In the cylindrically symmetric case, when the Abelian G_2 group acts orthogonally transitively, the metric can be written in the form of Eq. (3.41), but with the x^3-coordinate having closed orbits (Bronnikov, Santos and Wang, 2019). Then, the Einstein vacuum field equations reduce to the "cylindrical" Ernst equation (Chandrasekhar, 1986). Therefore, such isometries should exist between cylindrical spacetimes (Wang, 2003) and the interiors of black holes, too.

the solutions (4.23) reduce to

$$ds^2 = C_0(1+\eta)^2 \left(\frac{d\eta^2}{1-\eta^2} - \frac{d\mu^2}{1-\mu^2} \right) - \left(\frac{1-\eta}{1+\eta} \right)(dx^2)^2$$
$$- (1-\mu^2)(1+\eta)^2(dx^3)^2. \tag{6.2}$$

Introducing the new coordinates (Chandrasekhar and Xanthopoulos, 1986a, 1986b; Ferrari and Ibañez, 1988),

$$\eta \equiv \frac{r-m}{m}, \quad \mu \equiv \cos\theta, \quad x^2 \equiv t, \quad x^3 \equiv m\phi, \quad C_0 = m^2, \tag{6.3}$$

we find that the metric (6.2) takes the form,

$$ds^2 = \left(1 - \frac{2m}{r} \right) dt^2 - \left(1 - \frac{2m}{r} \right)^{-1} dr^2 - r^2(d\theta^2 + \sin\theta^2 d\phi^2), \tag{6.4}$$

which is precisely the form of the Schwarzschild black hole solution written in the spherically symmetric coordinates (t, r, θ, ϕ). Note that the focusing hypersurface $\eta = 1$ or $t = 0$ corresponds to the Killing-horizon $r = 2m$, in other words, now the focusing hypersurface is not singular, and instead it represents a horizon. This is consistent with the analysis presented between Eqs. (4.45) and (4.47). In particular, the choice of Eq. (6.1) satisfies the non-singular condition (4.47).

It should be noted that the above equivalence holds only locally, as one can see from Eq. (6.3), which tells us that it is the x^2-coordinate that is identical to the t-coordinate, while the space-like coordinate μ is identical to the angular coordinate ϕ, and the time-like coordinate η to r, as inside the black hole, r becomes time-like, and t is space-like.

On the other hand, from Eq. (4.28) we find that the choice of Eq. (6.1) yields,

$$m = n = 1, \tag{6.5}$$

for which the incoming gravitational waves consist of two parts, the impulsive and shock parts, as one can see from the analysis given between Eqs. (4.34) and (4.42). In particular, along the wavefronts $u = 0$ and $v = 0$, the two wave components Ψ_0 and Ψ_4 takes the forms,

$$\Psi_0 = -3H(v) + \delta(v), \quad \text{(across } v = 0\text{),}$$
$$\Psi_4 = -3H(u) + \delta(u), \quad \text{(across } u = 0\text{).} \tag{6.6}$$

Inside Region II (III), a right-hand (left-hand) moving gravitational shock wave is present, given, respectively, by

$$\Psi_0^{II}(v) = \frac{3}{(1+v)^2(1-v)}, \quad \text{(Region II)},$$

$$\Psi_4^{III}(u) = \frac{3}{(1+u)^2(1-u)}, \quad \text{(Region III)}. \tag{6.7}$$

It is equally remarkable that the following specification of the free parameters also leads to the internal solution of the Schwarzschild black hole. In fact, setting (Ferrari and Ibañez, 1988),

$$a = +\frac{1}{2}, \quad \delta_1 = 1, \quad \delta_2 = 0, \tag{6.8}$$

in the solutions given by Eq. (4.23), we find that

$$ds^2 = C_0(1-\eta)^2 \left(\frac{d\eta^2}{1-\eta^2} - \frac{d\mu^2}{1-\mu^2} \right) - \left(\frac{1+\eta}{1-\eta} \right) (dx^2)^2$$
$$- (1-\mu^2)(1-\eta)^2(dx^3)^2, \tag{6.9}$$

which can be obtained from the metric (6.2) by the replacement $\eta \to -\eta$. However, now the coordinate transformations (6.3) to the internal region of the Schwarzschild black hole solution (6.4) become,

$$\eta \equiv \frac{m-r}{m}, \quad \mu \equiv \cos\theta, \quad x^2 \equiv t, \quad x^3 \equiv m\phi, \quad C_0 = m^2, \tag{6.10}$$

that is, all of them are the same as these given by Eq. (6.3) except for $\eta \to -\eta$. It is this difference that now the focusing hypersurface $t = 0$ or $\eta = 1$ corresponds to $r = 0$, instead of $r = 2m$, as given in the last case. Therefore, in the present case the hypersurface $t = 0$ ($\eta = 1$) becomes a spacetime curvature singularity. Note that this is also consistent with the analysis given between Eqs. (4.45) and (4.47), and in particular the non-singular condition (4.47) is not satisfied.

From Eq. (4.28) we find that the choice of Eq. (6.8) also yields $m = 1 = n$, and along the wavefronts $u = 0$ and $v = 0$, the two gravitational wave components Ψ_0 and Ψ_4 take the forms,

$$\Psi_0 = 3H(v) + \delta(v), \quad \text{(across } v = 0\text{)},$$
$$\Psi_4 = 3H(u) + \delta(u), \quad \text{(across } u = 0\text{)}. \tag{6.11}$$

Inside Region II (III), a right-hand (left-hand) moving gravitational shock wave is present, but now given, respectively, by

$$\Psi_0^{II}(v) = \frac{3}{(1-v)^2(1+v)}, \quad \text{(Region II)},$$
$$\Psi_4^{III}(u) = \frac{3}{(1-u)^2(1+u)}, \quad \text{(Region III)}. \tag{6.12}$$

Comparing Eq. (6.6) with Eq. (6.11), we can see that the shock wave part changes the signs, while their impulsive parts remain the same. On the other hand, Eqs. (6.7) and (6.12) show that the shock waves along the hypersurfaces $u = 1$ and $v = 1$ become more singular in the latter, proportional to, respectively, $(1-u)^{-2}$ and $(1-v)^{-2}$, while in the former, they are proportional to, respectively, $(1-u)^{-1}$ and $(1-v)^{-1}$.

6.2.   The Kerr–NUT Black Hole

Chandrasekhar and Ferrari (1984) first showed that the Nutku–Halil solution (Nutku and Halil, 1977) is given by the Ernst potential [see Eq. (4.141)],

$$E = p\eta + iq\mu, \quad (p^2 + q^2 = 1), \tag{6.13}$$

where E is directly related to the metric coefficients V and W via the relations [see Eq. (3.44)],

$$Z \equiv \chi + iq_2, \quad E \equiv \frac{Z-1}{Z+1}, \quad \chi \equiv \frac{e^V}{\cosh W}, \quad q_2 \equiv \frac{\sinh W e^V}{\cosh W}. \tag{6.14}$$

The Nutku–Halil solution represents the collision of two pure gravitational impulsive waves with uncorrelated (non-collinear) polarizations. It is a direct generalization of the Khan–Penrose solution (Khan and Penrose, 1971), which represents the collision of two pure gravitational impulsive waves with correlated (collinear) polarizations, and the corresponding Ernst potential is given by Eq. (6.13) with $q = 0$.[2]

As first noticed by Chandrasekhar and Ferrari (1984), the Kerr solution (Kerr, 1963) also follows from the same Ernst potential (6.13) (Chandrasekhar, 1983). But, in the stationary case the Ernst potential is related to the metric coefficients through the function $\tilde{\Phi}$ and $\tilde{\Psi}$ (Chandrasekhar,

---

[2]Barrabés, Bressange and Hogan showed that the Khan–Penrose and Nutku–Halil solutions can be also obtained from Einstein's vacuum field equations as an initial value problem (Barrabés, Bressange and Hogan, 1999).

1983, Section 54),

$$Z = \tilde{\Psi} + i\tilde{\Phi}, \tag{6.15}$$

where $\tilde{\Phi}$ is the potential for q_2, defined similar to Q given by Eq. (4.8), and $\tilde{\Psi}$ similar to P defined by Eq. (4.10).

Therefore, to get something similar to the Kerr solution, we need first to solve Eq. (4.8) for q_2, and then read off χ from Eq. (4.10) with

$$P + iQ = \frac{1 + E^\dagger}{1 - E^\dagger}, \tag{6.16}$$

where $E^\dagger = p\eta + iq\mu$. In doing so, Chandrasekhar and Xanthopoulos (1986a, 1986b) found that

$$\chi = (1 - \eta^2)^{1/2}(1 - \mu^2)^{1/2}\frac{(1 - p\eta)^2 + q^2\mu^2}{1 - p^2\eta^2 - q^2\mu^2},$$

$$q_2 = \frac{2q}{1 + p}\frac{(1 - \eta)(p\mu^2 + p\eta + \mu^2 - 1)}{1 - p^2\eta^2 - q^2\mu^2}, \tag{6.17}$$

$$f = \frac{(1 - p\eta)^2 + q^2\mu^2}{\eta^2 - \mu^2}.$$

To extend the above solution to Regions I–III, one can use the Khan–Penrose substitutions, and then it can be shown that the solution represents the collision of two pure gravitational impulsive waves, each of which is accompanied by a gravitational shock wave (Chandrasekhar and Xanthopoulos, 1986a, 1986b).

In the interacting region (Region IV), setting (Chandrasekhar and Xanthopoulos, 1986a, 1986b)

$$\eta = \mp\frac{M - r}{\sqrt{M^2 - J^2}}, \quad \mu = \cos\theta,$$

$$t = M\left[x^2 - \frac{2q}{p(1 + p)}x^3\right], \quad \phi = \frac{M}{\sqrt{M^2 - J^2}}x^3, \tag{6.18}$$

$$p = \mp\frac{\sqrt{M^2 - J^2}}{M}, \quad q = \pm\frac{J}{M},$$

the solution can be cast in the form,

$$ds^2 = \frac{\rho^2 - 2Mr}{\rho^2}\left(dt + \frac{2JMr\sin^2\theta}{\rho^2 - 2Mr}d\phi\right)^2 - \frac{\rho^2}{\Delta}[dr^2 + \Delta d\theta^2]$$

$$- \frac{\Delta\rho^2\sin^2\theta}{\rho^2 - 2Mr}d\phi^2, \tag{6.19}$$

where

$$\Delta \equiv r^2 - 2Mr + J^2, \quad \rho \equiv \sqrt{r^2 + J^2 \cos^2 \theta}. \tag{6.20}$$

The above solution is exactly the Kerr solution written in the Boyer–Lindquist coordinates with M and J being the mass and angular momentum parameters (Chandrasekhar, 1983).

Since $-1 \leq \eta \leq +1$, from Eq. (6.18) we find that this implies

$$r_- \leq r \leq r_+, \quad (r_\pm = M \pm \sqrt{M^2 - J^2}), \tag{6.21}$$

that is, the interacting region of the two pure colliding plane gravitational waves is isometric to the ergo-sphere of the Kerr black hole, in which the time-like Killing vector ∂_t becomes space-like (in addition to the space-like Killing vector ∂_ϕ). As a result, in this region there are two space-like Killing vectors, which are consistent with the requirement of spacetimes with plane symmetry (Stephani *et al.*, 2009), although in the Kerr spacetime, one of them has a closed orbit [cf. Eq. (6.18)]. Therefore, such an equivalence can be only local.

On the other hand, setting $\eta = 1$, which corresponds to the focusing hypersurface $t = 0$, from Eq. (6.18) we find that

$$\overset{.}{r}(t = 0) = M \pm \sqrt{M^2 - J^2}. \tag{6.22}$$

Thus, if the "$-$" sign in Eq. (6.18) is chosen, the focusing surface corresponds to the null surface $r = r_+$, which is the location of the event horizon of the Kerr black hole, and when the "$+$" sign in Eq. (6.18) is chosen, the focusing surface corresponds to the null surface $r = r_-$, which is the location of the Cauchy horizon of the Kerr black hole. In each of the two cases, the spacetime is not singular at $t = 0$, and extensions beyond these surfaces are needed. Clearly, if we require the extension is analytical, then we shall get a global structure quite similar to that of the Kerr black hole. However, in the case $r(t = 0) = r_-$ the focusing hypersurface is a Cauchy horizon, which is not stable against small perturbations (Yurtsever, 1987, 1988a), so such extensions might not be needed in this case.

It is interesting to note that the solution (6.17) is precisely the two-soliton solution with

$$a = \frac{1}{2}, \quad \delta_1 = \delta_2 = 0, \quad \Rightarrow \quad n = m = 1, \tag{6.23}$$

given by Eqs. (4.142) and (4.143) by setting the NUT parameter l to zero.

When $l \neq 0$, the interacting region of the colliding plane gravitational wave is isometric to the interior of the Kerr–NUT solution. When $q = 0$, the solution reduces to that given by Eq. (4.145), in which the interacting region is isometric to the interior of the Taub–NUT solution, first studied by Ferrari and Ibañez (1988).

Finally, we note that in the current case we have $\alpha \equiv a + \delta_1 = 1/2$, which satisfies one of the conditions given in Eq. (4.134), and as a result the corresponding solution is not singular at the focusing surface $t = 0$. This is well consistent with the above analysis.

6.3. Other Black Holes

From the analysis given in the last two sections, we can see that, for any given stationary spacetime, M, if it contains at least one more space-like Killing vector, say, ∂_x, then this Killing vector will form a G_2 group together with the time-like Killing vector ∂_t. In such a space-time, if we further assume that a black hole exists, then by definition this time-like Killing vector ∂_t will become space-like inside the black hole (or at least inside a region of the black hole, such as that in the case of the Kerr solution considered in the last section). Then, in this region the two space-like Killing vectors can be (locally) identified with the two space-like Killing vectors ∂_{x^2} and ∂_{x^3} for the colliding plane wave space-times. Restricting oneself to this internal region, and then using the Khan–Penrose substitutions

$$u \to uH(u), \quad v \to vH(v), \tag{6.24}$$

one can obtain the spacetime in Regions I–III. Then, such resulted space-time will represent the collision of gravitational plane waves (possibly coupled with matter), provided that certain physical conditions are satisfied, such as the energy conditions (Hawking and Ellis, 1973), and satisfying matter field equations. Clearly, such an obtained solution is isometric to the interior of the black hole.

Along this vein, it can be shown that the interior of the Kerr–Newman charged black hole is isometric to the interacting region of two colliding gravitational plane waves, each of which is coupled with an electromagnetic wave (Chandrasekhar and Xanthopoulos, 1987a, 1987b). Along the same line, it was found that the interior of the Kerr–Newman–NUT solution (Sephani et al., 2009) is also isometric to an interacting region of two colliding gravitational plane waves coupled with electromagnetic plane waves (Gurtug and Halilsoy, 2009).

Similarly, Papacostas and Xanthopoulos (1989) found a five-parameter class of solutions, which describe the collision of plane-fronted impulsive gravitational and shock electromagnetic waves. Again, the interaction region of the two colliding waves is a locally known solution of the Einstein–Maxwell electrovacuum equations of Petrov type D, and the collision results in the formation of a Cauchy horizon.

In addition, Griffiths and Halburd (2007) showed that part of the C-metric spacetime inside the black hole can be also interpreted as the interaction region of two colliding plane waves with aligned linear (collinear) polarizations. The focusing surface $t = 0$ is not singular, and instead represents a Cauchy horizon. Note that the C-metric was first found by Weyl in 1917 (Weyl, 1917), and subsequently rediscovered by many authors. But, it was Kinnersley and Walker (1970) and Bonnor (1983) who first showed that its analytic extension represents a pair of black holes that accelerate away from each other due to the presence of a strut along the symmetry axis.

Chapter 7

Concluding Remarks

In this book, we study the collision and nonlinear interaction of two pure gravitational plane waves, or a gravitational plane wave with a matter wave, or two matter waves, within the framework of Einstein's theory of general relativity, by using the distribution theory, first introduced by Taub to general relativity in 1980 (Taub, 1980). The method is mathematically equal to Israel's junction conditions (Israel, 1966, 1967), when the singular hypersurface is time-like or space-like. When the surface is null, it is equivalent to the analysis of Barrabés (see Barrabés, 1989; Barrabés and Hogan, 2003). Thus, they can be considered as complementary to each other.

The advantage of the studies of spacetimes for colliding gravitational and/or matter plane waves is that they can be investigated analytically, thanks to the development of nonlinear differential equations in 1960s (Gardner, Green, Kruskal and Miura, 1967; Whitham, 1974), and applications to Einstein's general relativity in 1970s and 1980s (Belinsky and Verdaguer, 2001; Griffiths and Podolský, 2009; Stephani *et al.*, 2009), including the soliton technique of Blelinsky and Zakharov (1978, 1979), its generalization to Einstein–Maxwell equations (Alekseev, 1981), and the Bäcklund transformations (Harrison, 1978, 1980; Neugebauer, 1979). With these techniques, most of the well-known solutions were rediscovered, including the Kerr–Newman–NUT black hole solutions (Belinsky and Verdaguer, 2001).

In fact, such techniques can be applied to all the spacetimes with two orthogonal Killing vectors, including stationary axially symmetric spacetimes (Griffiths and Podolský, 2009; Stephani *et al.*, 2009), cylindrical spacetimes (Verdaguer, 1993; Bronnikov, Santos and Wang, 2019), and spacetimes with plane symmetry, studied in this book.

With the studies of exact solutions that represent colliding plane wave spacetimes, now it is clear that spacetime singularities are generically developed due to the nonlinear interaction of the two colliding plane waves, as shown explicitly in Chapters 4 and 5. Killing–Cauchy horizons can be formed in particular cases (Chandrasekhar and Xanthopoulos, 1986a), but they are not stable against small perturbations (Yurtsever, 1987; Clarke and Hayward, 1989; Griffiths, 2005), and are expected to be turned into real spacetime singularities.

One might think that this is due to the high symmetry of the spacetimes, and in particular because of the plane symmetry, the incoming waves always have infinitely large amount of energy. However, recently it was showed that, due to such nonlinear interactions, spacetime singularities can be still formed in the head-on collision of axisymmetric distributions of null particles even with finite energy (Pretorius and East, 2018).

This result is extremely encouraging, and motivating various interesting questions. One is regarding to the internal structure of black holes. As shown in Chapter 6, the interiors of the most well-known black holes are isometric to the interacting regions of two colliding plane waves. Is it possible to shed lights on the internal structure of black holes by studying the head-on collision of two gravitational and/or matter wave? What about cosmic censorship and hoop conjectures (Penrose, 1969; Thorne, 1972)? Can we also be able to say something about the critical phenomena of gravitational collapse in the threshold of black hole formation (Choptuik, 1993; see also the review articles, Wang, 2001; Gundlach and Martin-Garcia, 2007).

To be more realistic, one might like first to remove the assumption of uniform plane-fronted waves, so one can deal with the case in which incoming waves have finite energy, similar to what were done by Pretorius and East (2018), and more recently by Baumgarte, Gundlach and Hilditch (2019). Perturbative calculations of this kind already started in the early of 1990s by D'Eath and Payne (1992a, 1992b, 1992c), and such obtained resulted are quite consistent with these numerical ones (Pretorius and East, 2018). Recently, such studies have been also generalized to other cases, including that the background is an expanding universe (Centrella and Matzner, 1982; Alekseev, 2016).

The above works have been mainly related to some theoretical aspects of Einstein's theory. With entering the era of gravitational wave astronomy, one might like also to ask more observational ones, such as the detection of gravitational memory effects (Favata, 2010; Bieri, Garfinkle and Yunes, 2017), just briefly mentioned in Section 2.5, in addition to the more theoretical works of soft-graviton theorems (Hawking, Perry and Strominger, 2016, 2017; Strominger, 2017).

In Chapter 3, we find that, due to the nonlinear interaction, the polarization of a colliding gravitational plane wave can be also get changed, a precise analog of the electromagnetic Faraday rotation (Faraday, 1846a, 1846b), but now with the oppositely moving wave, either gravitational or matter, as both the magnetic field and medium. Then, a natural question raises: is it possible in the future to observe such phenomena? The answer must be positive, but the real question is when?

With the detections of gravitational waves by LIGO/Virgo (Abbott *et al.*, 2019), which will be jointed soon by KAGRA (Akutsu *et al.*, 2019), and the direct detection of the supermassive M87 black hole by EHT (Akiyama *et al.*, 2019), two unprecedented new windows to study the strong gravitational field regime have just opened, and indeed we are entering the epoch of exploring the nonlinear nature of Einstein's general relativity, the characteristics that bare stark contrast to Newtonian theory, not only theoretically but also experimentally! All these just mark the beginning of a new era, in which *"the possibilities for discovery are as rich and boundless as they have been with light-based astronomy."*

Bibliography

Abbott, B.P. *et al.* (LIGO Scientific Collaboration and Virgo Collaboration). (2016). Observation of gravitational waves from a binary black hole merger, *Phys. Rev. Lett.* **116**, 061102.

Abbott, B.P. *et al.* (LIGO Scientific Collaboration and Virgo Collaboration). (2017). GW170817: Observation of gravitational waves from a binary neutron star inspiral, *Phys. Rev. Lett.* **119**, 161101.

Abbott, B.P. *et al.* (LIGO Scientific Collaboration and Virgo Collaboration). (2019). GWTC-1: A gravitational-wave transient catalog of compact binary mergers observed by LIGO and Virgo during the first and second observing runs, *Phys. Rev. X* **9**, 031040.

Ade, P.A.R. *et al.* (2016). Planck 2015 results. XIII. Cosmological parameters, *Astron. Astrophys.* **594**, A13.

Aichelburg, P.C. and Sexl, R.U. (1971). On the gravitational field of a massless particle, *Gen. Relativ. Grav.* **2**, 303.

Akiyama, K. *et al.* (The Event Horizon Telescope Collaboration). (2019). First M87 event horizon telescope results. I. The shadow of the supermassive black hole, *Astrophys. J. Lett.* **875**, L1.

Akutsu, T. *et al.* (KAGRA Collaboration). (2019). KAGRA: 2.5 generation interferometric gravitational wave detector, *Nature Astronom.* **3**, 35.

Alekseev, G.A. (1981). N-soliton solutions of Einstein–Maxwell equations, *JETP Lett.* **32**, 277.

Alekseev, G.A. (2016). Collision of strong gravitational and electromagnetic waves in the expanding universe, *Phys. Rev. D* **93**, 061501(R).

Alekseev, G.A. and Griffiths, J.B. (2001). Solving the characteristic initial value problem for colliding plane gravitational and electromagnetic waves, *Phys. Rev. Lett.* **87**, 221101.

Alekseev, G.A. and Griffiths, J.B. (2004). Collision of plane gravitational and electromagnetic waves in a Minkowski background: Solution of the characteristic initial value problem, *Class. Quantum Grav.* **21**, 5623.

Álvarez-Gaumé, L. *et al.* (2009). Critical formation of trapped surfaces in the collision of gravitational shock waves, *J. High Energy Phys.* **02**, 009.

Apostolatos, T.A. and Thorne, K.A. (1992). Rotation halts cylindrical, relativistic gravitational collapse, *Phys. Rev. D* **46**, 2435.

Arkani-Hamed, N. Dimopoulos, S. and Dvali, G. (1998). The hierarchy problem and new dimensions at a millimeter, *Phys. Lett. B* **429**, 263.

Arkani-Hamed, N. Dimopoulos, S. and Dvali, G. (1999). Phenomenology, astrophysics, and cosmology of theories with submillimeter dimensions and TeV scale quantum gravity, *Phys. Rev. D* **59**, 086004.

Babala, D. (1987). Collision of a gravitational impulsive wave with a shell of null dust, *Class. Quantum Grav.* **4**, L89.

Baldwin, O.R. and Jeffery, G.B. (1926). The relativity theory of plane waves, *Proc. R. Soc. A* **111**, 95.

Barrabés, C. (1989). Singular hypersurfaces in general relativity: A unified description, *Class. Quantum Grav.* **6**, 581.

Barrabés, C. Bressange, G.F. and Hogan, P.A. (1999). Colliding plane impulsive gravitational waves, *Prog. Theor. Phys.* **102**, 1085.

Barrabés, C. and Hogan, P.A. (2003). *Singular Null Hypersurfaces in General Relativity: Light-like Signals from Violent Astrophysical Events* (World Scientific, Singapore).

Barrabés, C. and Hogan, P.A. (2014). Generating a cosmological constant with gravitational waves, *Gen. Relativ. Grav.* **46**, 1635.

Barrabés, C. and Hogan, P.A. (2015). Colliding impulsive gravitational waves and a cosmological constant, *Phys. Rev. D* **92**, 044032.

Barrabés, C. Gramain A. Lesigne E. and Letelier P.S. (1992). Geometric inequalities and the hoop conjecture, *Class. Quantum Grav.* **9**, L105.

Barrabés, C. Israel W. and Letelier P.S. (1991). Analytic models of nonspherical collapse, cosmic censorship and the hoop conjecture, *Phys. Lett. A* **160**, 41.

Basov, N.G. (1979). *Problems in General Theory of Relativity and Theory of Group Representations.* The Lebedev Physics Institute Series, Vol. 96 (Springa US) (Translated from Russian by A Mason, New York).

Baumgarte, T.W. Gundlach, C. and Hilditch, D. (2019). Critical phenomena in the gravitational collapse of electromagnetic waves, *Phys. Rev. Lett.* **123**, 171103.

Baumgarte, T.W. and Shapiro, S.L. (2010). *Numerical Relativity: Solving Einstein's Equations on the Computer* (Cambridge University Press, Cambridge).

Belinsky, V.A. (1979). One-soliton cosmological waves, *Sov. Phys. JETP* **50**, 623.

Belinsky, V.A. (1980). L–A pair of a system of coupled equations of the gravitational and electromagnetic fields, *JETP Lett.* **30**, 29.

Belinsky, V.A. and Francaviglia, M, (1982). Solitonic gravitational waves in Bianchi II cosmologies. 1. The general framework, *Gen. Relativ. Grav.* **14**, 213.

Belinsky, V.A. and Ruffini, R. (1980). On axially symmetric soliton solutions of the coupled scalar-vector-tensor equations in general relativity, *Phys. Lett. B* **89**, 195.

Belinsky, V.A. and Verdaguer, E. (2001). *Gravitational Solitons* (Cambridge University Press, Cambridge).

Belinsky, V.A. and Zakharov, V.E. (1978). Integration of the Einstein equations by means of the inverse scattering problem technique and construction of exact soliton solutions, *Sov. Phys. JETP* **48**, 985.

Belinsky, V.A. and Zakharov, V.E. (1979). Stationary gravitational solitons with axial symmetry, *Sov. Phys. JETP* **50**, 1.

Bell, P. and Szekeres, P. (1974). Interacting electromagnetic shock waves in general relativity, *Gen. Relativ. Grav.* **5**, 275.

Bennett, C.L. Banday, A. Gorski, K.M. Hinshaw, G. Jackson, P. Keegstra, P. Kogut, A. Smoot, G.F. Wilkinson, D.T. and Wright, E.L. (1996). Four-year COBE DMR cosmic microwave background observations: Maps and basic results, *Astrophys. J.* **464**, L1.

Berezin, V.A. Kuzmin, V.A. and Tkachev, I. (1987). Dynamics of bubbles in general relativity, *Phys. Rev. D* **36**, 2919.

Bieri, L. Garfinkle, D. and Yunes, N. (2017). Gravitational waves and their mathematics, *AMS Notices* **64**, 07.

Blanchet, L. and Damour, T. (1992). Hereditary effects in gravitational radiation, *Phys. Rev. D* **46**, 4304.

Bondi, H. (1957). Plane gravitational waves in general relativity, *Nature (London)* **179**, 1072.

Bondi, H. and Pirani, F.A.E. (1989). Gravitational waves in general relativity XIII. caustic property of plane waves, *Proc. R Soc. Lond. A* **421**, 395.

Bondi, H. Pirani, F.A.E. and Robinson, I. (1959). Gravitational waves in general relativity III. Exact plane waves, *Proc. R Soc. Lond. A* **251**, 519.

Bonnor, W.B. (1983). Physical interpretation of vacuum solutions of Einstein's equations. Part I. Time-independent solutions, *Gen. Relativ. Grav.* **15**, 535.

Bonnor, W.B. and Vichers, P.A. (1981). Junction conditions in general relativity, *Gen. Relativ. Grav.* **13**, 29.

Boulware, D.G. (1973). Naked singularities, thin shells, and the Reissner-Norstrom metric, *Phys. Rev. D* **8**, 2363.

Braginsky, V.P. and Grishchuk, L.P. (1985). Kinematic resonance and memory effect in free-mass gravitational antennas, *Zh. Eksp. Teor. Fiz.* **89**, 744 [*Sov. Phys. JETP* **62**, 427].

Braginsky, V.P. and Thorne, K.P. (1987). Gravitational-wave bursts with memory and experimental prospects, *Nature* **327**, 123.

Brandenberger, R.H. (1985). Quantum field theory methods and inflationary universe models, *Rev. Mod. Phys.* **57**, 1.

Braun, V. and Ovrut, B.A. (2006). Stabilizing moduli with a positive cosmological constant in heterotic M-theory, *J. High Energy Phys.*, **07**, 035.

Bretón, N. García, A. Macías, A. and Yáñez, G. (1998). Colliding plane waves in terms of Jacobi functions, *J. Math. Phys.* **39**, 6051.

Bretón, N. Matos, T. and García, A. (1996). Colliding plane waves in Einstein–Maxwell-dilaton fields, *Phys. Rev. D* **53**, 1868.

Brinkman. H.W. (1923). On riemann spaces conformal to einstein spaces, *Proc. Natl. Acad. Sci. USA* **9**, 1.

Bronnikov, K.A. (1980). Gravitational and sound waves in stiff matter, *J. Phys. A: Math. Gen.* **13**, 3455.

Bronnikov, K.A. Santos, N.O. and Wang, A. (2019). Cylindrical systems in general relativity, preprint, arXiv:1901.06561 [*Class. Quantum Grav.* in press (2020)].

Bruckman, W. (1986). Stationary axially symmetric exterior solutions in the five-dimensional representation of the Brans–Dicke–Jordan theory of gravitation, *Phys. Rev. D* **34**, 2990.

Bruckman, W. (1987). Exact axially symmetric stationary solutions of the Kaluza–Klein–Jordan–Thiry theory, *Phys. Rev. D* **36**, 3674.

Campbell, S.J. and Wainwright, J. (1977). Algebraic computing and the Newman-Penrose formalism in general relativity, *Gen. Relativ. Grav.* **8**, 987.

Carot, J. and Verdaguer, E. (1989). Generalised soliton solutions of the Weyl class, *Class. Quantum Grav.* **6**, 845.

Carr, B.J. and Verdaguer, E. (1984). Soliton solutions and cosmological gravitational waves, *Phys. Rev. D* **28**, 2995.

Carroll, S. (2004). *Spacetime and Geometry: An Introduction to General Relativity* (Addison-Wesley, New York).

Centrella, J. and Matzner, R.A. (1982). Colliding gravitational waves in expanding cosmologies, *Phys. Rev. D* **25**, 930.

Cespedes, J. and Verdaguer, E. (1987). Gravitational wave pulse and soliton wave collision, *Phys. Rev. D* **36**, 2259.

Challifour, J.L. (1972). *Generalized Functions and Fourier Analysis* (W. A. Benjamin, Reading, MA).

Chan, R. da Silva, M.F.A. da Rocha, J.F. and Wang, A. (2011). Radiating gravastars, *J. Cosmol. Astropart. Phys.* **10**, 013.

Chandrasekhar, S. (1983). *The Mathematical Theory of Black Holes* (Clarendon Press. Oxford).

Chandrasekhar, S. (1986). Cylindrical waves in general relativity, *Proc. R. Soc. Lond. A* **408**, 209.

Chandrasekhar, S. and Ferrari, V. (1984). On the Nutku–Halil solution for colliding impulsive gravitational waves, *Proc. R. Soc. Lond. A* **396**, 55.

Chandrasekhar, S. and Xanthopoulos, B.C. (1985a). On colliding waves in the Einstein-Maxwell theory, *Proc. R. Soc. Lond. A* **398**, 223.

Chandrasekhar, S. and Xanthopoulos, B.C. (1985b). On the collision of impulsive gravitational waves when coupled with fluid motions, *Proc. R. Soc. Lond. A* **402**, 37.

Chandrasekhar, S. and Xanthopoulos, B.C. (1985c). Some exact solutions of gravitational waves when coupled with fluid motions, *Proc. R. Soc. Lond. A* **402**, 205.

Chandrasekhar, S. and Xanthopoulos, B.C. (1986a). A new type of singularity created by colliding gravitational waves, *Proc. R. Soc. Lond. A* **408**, 175.

Chandrasekhar, S. and Xanthopoulos, B.C. (1986b). On the collision of impulsive gravitational waves when coupled with null dust, *Proc. R. Soc. Lond. A* **403**, 189.

Chandrasekhar, S. and Xanthopoulos, B.C. (1987a). The effect of sources on horizons that may develop when plane gravitational waves collide, *Proc. R. Soc. Lond. A* **414**, 1.

Chandrasekhar, S. and Xanthopoulos, B.C. (1987b). On colliding waves that develop time-like singularities: a new class of solutions of the Einstein - Maxwell equations, *Proc. R. Soc. Lond. A* **410**, 311.

Chandrasekhar, S. and Xanthopoulos, B.C. (1988). A perturbation analysis of the Bell–Szekeres space-time, *Proc. R. Soc. Lond. A* **420**, 93.

Chen, B. (2004). Colliding plane wave solutions in string theory revisited, *J. High Energy Phys.* **12**, 016.

Chen, B. Chu, C.S. Furuta, K. and Lin, F.L. (2004). Colliding plane waves string theory, *J. High Energy Phys.* **02**, 020.

Chen, B. Zhang, J.F. (2005). Two-flux colliding plane waves in string theory, *Commun. Theor. Phys.* **44**, 463.

Chernoff, D.F. Flanagan E.E. and Wardell, B. (2018). Gravitational back-reaction on a cosmic string: Formalism, preprint, arXiv:1808.08631.

Chernoff, D.F. and Tye, S.H.H. (2018). Detection of low tension cosmic super-strings, *J. Cosmol. Astropart. Phys.* **05**, 002.

Chesler, P.M. (2015). Colliding shock waves and hydrodynamics in small systems, *Phys. Rev. Lett.* **115**, 241602.

Chesler, P.M. and Yaffe, L.G. (2011). Holography and colliding gravitational shock waves in asymptotically AdS_5 spacetime, *Phys. Rev. Lett.* **106**, 021601.

Choptuik, M.W. (1993). Universality and scaling in gravitational collapse of a massless scalar field, *Phys. Rev. Lett.* **70**, 9.

Christodoulou, D. (1991). Nonlinear nature of gravitation and gravitational wave experiments, *Phys. Rev. Lett.* **67**, 1486.

Clarke, C.J.S. and Dray, T. (1987). Junction conditions for null hypersurfaces, *Class. Quantum Grav.* **4**, 265.

Clarke, C.J.S. and Hayward, S.A. (1989). The global structure of the Bell–Szekeres solution, *Class. Quantum Grav.* **6**, 615.

Connors, P.A. Piran, T. and Stark, R.F. (1980). Polarization features of X-ray radiation emitted near block hole, *Astrophys. J.* **235**, 224.

Connors, P.A. and Stark, R.F. (1977). Observable gravitational effects on polarised radiation coming from near a block hole, *Nature London* **269**, 128.

Copeland, E.J. Myers, R.C. and Polchinski, J. (2004). Cosmic F and D strings, *J. High Energy Phys.* **06**, 013.

Cosgrove, C.M. (1980). Relationships between the group-theoretic and soliton-theoretic techniques for generating stationary axisymmetric gravitational solutions, *J. Math. Phys.* **21**, 2417.

Cosgrove, C.M. (1981). Backlund transformations in the Hauser–Ernst formalism for stationary axisymmetric space-times, *J. Math. Phys.* **22**, 2624.

Cosgrove, C.M. (1982). Relationship between the inverse scattering techniques of Belinskii–Zakharov and Hauser–Ernst in general relativity, *J. Math. Phys.* **23**, 615.

Darmois, G. (1927). *Les équations de la gravitation einstetntenne*, Mémori al des Sciences, Mathématiques XXV (Gauthier-Villars. Paris).

Dautcourt, G. (1964). Wechselwirkungsfreie Gravitationsstoßwellen erster Ordnung, *Math. Nachr.* **27**, 277.

Davies, T.M. (1976a). A simple application of the Newman-Penrose spin coefficients formalism (I), *Int. J. Theor. Phys.* **15**, 315.

Davies, T.M. (1976b). A simple application of the Newman -Penrose spin coefficient formalism (II), *Int. J. Theor. Phys.* **15**, 319.

Da Silva, M.F.A. Herrera, L. Santos, N.O. and Wang, A. (2002). Rotating cylindrical shell source for Lewis spacetime, *Class. Quantum Grav.* **19**, 3809.

D'Eath, P.D. and Payne, P.N. (1992a). Gravitational radiation in black-hole collisions at the speed of light. I. Perturbation treatment of the axisymmetric collision, *Phys. Rev. D* **46**, 658.

D'Eath, P.D. and Payne, P.N. (1992b). Gravitational radiation in black-hole collisions at the speed of light. II. Reduction to two independent variables and calculation of the second-order news function, *Phys. Rev. D* **46**, 675.

D'Eath, P.D. and Payne, P.N. (1992c). Gravitational radiation in black-hole collisions at the speed of light. III. Results and conclusions, *Phys. Rev. D* **46**, 694.

De Leon, J.P. (2004). Brane-world models emerging from collisions of plane waves in 5D, *Gen. Relativ. Grav.* **36**, 923.

Deser, S. and Ford, K.W. (1964). *Lectures on General Relativity* Brandeis Summer Institute in Theoretical Physics, Vol. 1. (Prentice-Hall, Englewood Cliffs).

Devin, M. Ali, T. Cleaver, G. Wang, A. and Wu, Q. (2009). Branes in the $M_D \times M_{d+} \times M_{d-}$ Compactification of type II string on S^1/Z_2 and their cosmological applications, *J. High Energy Phys.* **10**, 095.

DeWitt, C. and DeWitt, B.S. (1964). *Relativity, Groups, and Topology* (Gordon and Breach, New York).

DeWitt, C. Schatzman, E. and Veron, P. (1967). *High Energy Astrophysics*, Vol. III (Gordon and Breach, New York).

DeWitt, C. and Wheeler, J.A. (1968). *Battelle Rencontres: 1967 Lectures in Mathematics and Physics* (W. A. Benjamin, New York).

Diaz, M. Gleiser, R.J. and Pullin, J.A. (1987). Solitonic perturbations of perfect fluid Friedmann-Robertson-Walker cosmological models, *Class. Quantum Grav.* **4**, 123.

Diaz, M. Gleiser, R.J. and Pullin, J.A. (1988). Solitonic solutions to the Kaluza-Klein-Jordan formalism as cosmological models in general relativity, *J. Math. Phys.* **29**, 169.

Ding, X. Q. and Ding, Y. (2005). *Hermite Expansion and Generalized Functions* (Huazhong Normal University Press, 2005).

D'Inverno, R. (2003). *Introducing Einstein's Relativity* (Clarendon Press, Oxford).

Dray, T. and 't Hooft, G. (1985a). The gravitational shock wave of a massless particle, *Nucl. Phys. B* **253**, 173.

Dray, T. and 't Hooft, G. (1985b). The effect of spherical shells of matter on the Schwarzschild black hole, *Commun. Math. Phys.* **99**, 613.

Dray, T. and 't Hooft, G. (1986). The gravitational effect of colliding planar shells of matter, *Class. Quantum Grav.* **3**, 825.

Dueñas-Vidal, A. and Vázquez-Mozo, M.A. (2010). Colliding AdS gravitational shock waves in various dimensions and holography, *J. High Energy Phys.* **07**, 021.

Dueñas-Vidal, A. and Vázquez-Mozo, M.A. (2012). A Note on the collision of Reissner–Nordström gravitational shock waves in AdS, *Phys. Lett. B* **713**, 500.

Dvali, G. and Vilenkin, A. (2004). Formation and evolution of cosmic D strings, *J. Cosmol. Astropart. Phys.* **03**, 010.

Echeverria F. (1993). Gravitational collapse of an infinite, cylindrical dust shell, *Phys. Rev. D* **47**, 2271.

Economou, A. (1988). Ph.D. dissertation, Physics Department, Ioannina University, Greece.

Economou, A. and Tsoubelis, D. (1988a). Rotating cosmic strings and gravitational soliton waves, *Phys. Rev. D* **38**, 498.

Economou, A. and Tsoubelis, D. (1988b). Interaction of cosmic strings with gravitational waves: A new class of exact solutions, *Phys. Rev. Lett.* **61**, 2046.

Economou, A. and Tsoubelis, D. (1989). Multiple-soliton solutions of Einstein's equations, *J. Math. Phys.* **30**, 1562.

Ehlers, J. and Kundt, W. (1962). Exact solutions of the gravitational field equations in *Gravitation: An Introduction to Current Research*, ed. Witten, L. (John Wiley & Sons), pp. 49–101.

Einstein, A. (1905). Zur Elektrodynamik bewegter Körper, *Ann. Phys.* **17**, 891.

Einstein, A. (1915a). Zur Allgemeinen Relativitätstheorie, *Preuss. Akad. Wiss. Berlin, Sitzber.* 778.

Einstein, A. (1915b). Der Feldgleichungen der Gravitation, *Preuss. Akad. Wiss. Berlin, Sitzber.* 844.

Einstein, A. (1916). Die Grundlage der allgemeinen Relativitätstheorie, *Ann. Phys.* **49**, 769.

Ellis, G.F.R. and Schmidt, B.G. (1977). *Singular space-times*, Gen. Relativ. Grav. **8**, 915.

Ernst, F.J. (1968a). New formulation of the axially symmetric gravitational field problem, *Phys. Rev.* **167**, 1175.

Ernst, F.J. (1968b). New formulation of the axially symmetric gravitational field problem. II, *Phys. Rev*, **168**, 1415.

Ernst, F.J. Garcia, A.D. and Hauser, I. (1987a). Colliding gravitational plane waves with non-collinear polarization. I, *J. Math. Phys.* **28**, 2155.

Ernst, F.J. Garcia, A.D. and Hauser, I. (1987b). Colliding gravitational plane waves with non-collinear polarization. II, *J. Math. Phys.* **28**, 2951.

Ernst, F.J. Garcia, A.D. and Hauser, I. (1988). Colliding gravitational plane waves with non-collinear polarization. III, *J. Math. Phys.* **29**, 681.

Faraday, M. (1846a). I. Experimental researches in electricity — Nineteenth series. *Philos. Trans.* **136**.

Faraday, M. (1846b). *Ann. Phys.* **68**, 105.

Farnsworth, D. Fink, J. Porter, J. and Thompson, A. (1972). Methods of local and global differential geometry in general relativity, in *Proceedings of the Regional Conference on Relativity*, University of Pittsburgh, Pittsburgh, Pennsylvania, July 13–17, 1970 (Springer-Verlag).

Favata, M. (2010). The gravitational–wave memory effect, *Class. Quantum Grav.* **27**, 084036.

Feinstein, A. (2002). Penrose limits, the colliding plane-wave problem and the classical string backgrounds, *Class. Quantum Grav.* **19**, 5353.

Feinstein, A. and Charach, C. (1986). Gravitational solitons and spatial flatness, *Class. Quantum Grav.* **3**, L5.

Feinstein, A. and Ibañez, J. (1989). Curvature-singularity-free solutions for colliding plane gravitational waves with broken $u - v$ symmetry, *Phys. Rev. D* **39**, 470.

Feinstein, A. MacCallum, M.A.H. and Senovilla, J.M.M. (1989). On the ambiguous evolution and the production of matter in space-times with colliding waves, *Class. Quantum Grav.* **6**, L217.

Feinstein, A. and Senovilla, J.M.M. (1989). Collision between variably polarized plane gravitational wave and a shell of null matter, *Phys. Lett. A* **138**, 102.

Ferrari, V. (1988). Focusing process in the collision of gravitational plane waves, *Phys. Rev. D* **37**, 3061.

Ferrari, V. and Ibañez, J. (1987a). On the collision of gravitational plane waves: a class of soliton solutions, *Gen. Relativ. Grav.* **19**, 405.

Ferrari, V. and Ibañez, J. (1987b). A new exact solution for colliding gravitational plane waves, *Gen. Relativ. Grav.* **19**, 383.

Ferrari, V. and Ibañez, J. (1988). Type-D solutions describing the collision of plane-fronted gravitational waves, *Proc. R. Soc. Lond. A* **417**, 417.

Ferrari, V. and Ibañez, J. (1989a). On the gravitational interaction of null fields, *Class. Quantum Grav.* **6**, 1805.

Ferrari, V. and Ibañez, J. (1989b). Gravitational interaction of massless particles, *Phys. Lett. A* **141**, 233.

Ferrari, V. Ibañez, J. and Bruni, M. (1987a). Colliding gravitational waves with non-collinear polarization: A class of soliton solutions, *Phys. Lett. A* **122**, 459.

Ferrari, V. Ibañez, J. and Bruni, M. (1987b). Colliding plane gravitational waves: A class of non-diagonal soliton solutions, *Phys. Rev. D* **36**, 1053.

Frolov, V.P. (1979). The Newman–Penrose method in the theory of general relativity, in *Problems in the General Theory of Relativity and Theory of Group Representations*, ed. Basov, N. G. and translated from Russian by Mason A. (Proceedings (Trudy) of the P. N. Lebedev Physics Institute, Vol. 96).

Garcá-Díaz, A. (1988). Colliding-wave generalizations of the Nutku–Halil metric in the Einstein–Maxwell theory, *Phys. Rev. Lett.* **61**, 507.

Garcá-Díaz, A. (1989). Colliding wave generalization of the Ferrari–Ibañez solution in the Einstein–Maxwell theory, *Phys. Lett. A* **138**, 370.

Gardner, C.S. Greene, J.M. Kruskal, M.D. and Miura, R.M. (1967). Method for solving the Korteweg-deVries equation, *Phys. Rev. Lett.* **19**, 1095.

Gelfand, I.M. and Shilov, G.E. (1964). *Generalized Functions. Vol. I: Properties and Operations* (Academic Press, New York, London).

Geroch, R. (1971). A method for generating solutions of Einstein's equations, *J. Math. Phys.* **12**, 918.

Geroch, R. (1972). A method for generating solutions of Einstein's equations II, *J. Math. Phys.* **13**, 394.

Goldberger, W.D. and Wise, M.B. (1999). Modulus stabilization with bulk fields, *Phys. Rev. Lett.* **83**, 4922.

Gong, Y.-G. Wang, A. and Wu, Q. (2008). Cosmological constant and late transient acceleration of the universe in the Horava–Witten heterotic M-theory on S^1/Z_2, *Phys. Lett. B* **663**, 147.

Gray, J. Lukas, A. and Ovrut, B. (2007). Flux, gaugino condensation, and antibranes in heterotic M-theory, *Phys. Rev. D* **76**, 126012.

Greenwald, J. Lenells, J. Satheeshkumar, V.H. and Wang, A. (2013). Gravitational collapse in Hořava–Lifshitz theory, *Phys. Rev. D* **88**, 024044.

Griffiths, J.B. (1976a). Colliding neutrino fields in general relativity, *J. Phys. A: Math. Gen.* **9**, 45.

Griffiths, J.B. (1976b). The collision of plane waves in general relativity, *Ann. Phys.* **102**, 388.

Griffiths, J.B. (1980). A problem with classical neutrino fields in general relativity, *Nukleonika*, **25**, 1415.

Griffiths, J.B. (1981). Neutrino fields in Einstein–Cartan theory, *Gen. Relativ. Grav.* **13**, 227.

Griffiths, J.B. (1985). On the Bell–Szekeres solution for colliding electromagnetic waves, in *Galaxies, Axisymmetric Systems and Relativity. Essays Presented to W.B. Bonnor on his 65th Birthday*, ed. MacCallum, M.A.H. (Cambridge University Press, Cambridge), p. 199.

Griffiths, J.B. (1987). Colliding plane gravitational waves, *Class. Quantum Grav.* **4**, 957.

Griffiths, J.B. (1990). Colliding plane gravito-electromagnetic waves, *Int. J. Theor. Phys.* **29**, 173.

Griffiths, J.B. (1991). *Colliding Plane Waves in General Relativity* (Clarendon Press, Oxford). Republished by Dover Publications, Inc. (New York, 2016).

Griffiths, J.B. (2005). The stability of Killing–Cauchy horizons in colliding plane wave space-times, *Gen. Relativ. Grav.* **37**, 1119.

Griffiths, J.B. and Halburd, R.G. (2007). The C-metric as a colliding plane wave spacetime, *Class. Quantum. Grav.* **24**, 1049.

Griffiths, J.B. and Podolsky, J. (2009). *Exact Space-Times in Einstein's General Relativity* (Cambridge University Press, Cambridge).

Griffiths, J.B. and Santano-Roco, M. (2002). The characteristic initial value problem for colliding plane waves: The linear case, *Class. Quantum Grav.*, **19**, 4273.

Grishchuk, L.P. and Polnarev, A.G. (1989). Gravitational wave pulses with velocity memory, *Sov. Phys. JETP*, **69** (1989) [*Zh. Eksp. Teor. Fiz.* **96**, 1153].

Grumiller, D. and Romatschke, P. (2008). On the collision of two shock waves in AdS_5, *J. High Energy Phys.* **08**, 027.

Gundlach, C. and Martin-Garcia, J.M. (2007). Critical phenomena in gravitational collapse, *Living Rev. Relativ.* **10**, 5.

Gürses, M. and Halilsoy, M. (1982). Interacting superposed electromagnetic shock plane waves in general relativity, *Lett. Nuovo Cimento*, **34**, 588.

Gürses, M. Kahya, E.O. and Karasu, A. (2002). Higher dimensional metrics of colliding gravitational plane waves, *Phys. Rev. D* **66**, 024029.

Gürses, M. and Sermutlu, E. (1995). Colliding gravitational plane waves in dilaton gravity, *Phys. Rev. D* **52**, 809.

Gurtug, O. and Halilsoy, M. (2000). Horizon instability in the cross polarized Bell–Szekeres spacetime, preprint, arXiv:gr-qc/000.6038.

Gurtug, O. and Halilsoy, M. (2009). Effect of NUT parameter on the analytic extension of the Cauchy horizon that develop in colliding wave spacetimes, *Int. J. Mod. Phys. A* **24**, 3171.

Gurtug, O. Halilsoy, M. and Sakalli, I. (2003). New singular and nonsingular colliding–wave solutions in Einstein-Maxwell-scalar theory, *Gen. Relativ. Grav.* **35**, 2159.

Gutperle, M. and Pioline, B. (2003). Type-IIB Colliding plane waves, *J. High Energy Phys.* **09**, 061.

Halli, M. (1979). Colliding plane gravitational waves, *J. Math. Phys.* **20**, 120.

Halilsoy, M. (1988a). Distinct family of colliding gravitational waves in general relativity, *Phys. Rev. D* **38**, 2979.

Halilsoy, M. (1988b). Colliding electromagnetic shock waves in general relativity, *Phys. Rev. D* **37**, 2121.

Halilsoy, M. (1989a). Colliding superposed waves in the Einstein–Maxwell theory, *Phys. Rev. D* **39**, 2172.

Halilsoy, M. (1989b). Colliding superposed waves in the Einstein–Maxwell theory, *Phys. Rev. D* **39**, 2172.

Halilsoy, M. Mazharimousavi, S.H. and Gurtug, O. (2014). Emergent cosmological constant from colliding electromagnetic waves, *J. Cosmol. Astropart. Phys.* **11**, 010.

Halilsory, M. and Sakalli, I. (2003). Collision of electromagnetic shock waves coupled with axion waves: An example, *Class. Quantum Grav.* **20**, 1417.

Harrison, B.K. (1978). Bäcklund transformation for the Ernst equation of general relativity, *Phys. Rev. Lett.* **41**, 1197.

Harrison, B.K. (1980). New large family of vacuum solutions of the equations of general relativity, *Phys. Rev. D* **21**, 1695.

Harte, A.I. (2013). Strong lensing, plane gravitational waves and transient flashes, *Class. Quantum Grav.* **30**, 075011.

Hassan, Z. Feinstein, A. and Manko, V. (1990). Asymmetric collision between plane gravitational waves with variably polarization, *Class. Quantum Grav.* **7**, L109.

Hauser, I. and Ernst, F.J. (1979a). Integral equation method for effecting Kinnersley–Chitre transformations, *Phys. Rev. D* **20**, 362.

Hauser, I. and Ernst, F.J. (1979b). Integral equation method for effecting Kinnersley-Chitre transformations. II, *Phys. Rev. D* **20**, 1783.

Hauser, I. and Ernst, F.J. (1980a). A homogeneous Hilbert problem for the Kinnersely–Chitre transformations, *J. Math. Phys.* **21**, 1126.

Hauser, I. and Ernst, F.J. (1980b). A homogeneous Hilbert problem for the Kinnersely–Chitre transformations of electrovac space-times, *J. Math. Phys.* **21**, 1418.

Hauser, I. and Ernst, F.J. (1981). Proof of a Geroch conjecture, *J. Math. Phys.* **22**, 1051.

Hauser, I. and Ernst, F.J. (1989a). Initial value problem for colliding plane gravitational waves. I, *J. Math. Phys.* **30**, 872.

Hauser, I. and Ernst, F.J. (1989b). Initial value problem for colliding plane gravitational waves. II, *J. Math. Phys.* **30**, 2322.

Hauser, I. and Ernst, F.J. (1990). Initial value problem for colliding plane gravitational waves. III, *J. Math. Phys.* **31**, 871.

Hawking, S.W. (1975). Particles creation by black holes, *Commun. Math. Phys.* **43**, 199.

Hawking, S.W. and Eillis, G.F.R. (1973). *The Large Scale Structure of Space-Time*, Vol. 1 (Cambridge University Press, Cambridge).

Hawking, S. W. Perry, M.J. and Strominger, A. (2016). Soft hair on black holes, *Phys. Rev. Lett.* **116**, 231301.

Hawking, S. W. Perry, M. J. and Strominger, A. (2017). Superrotation charge and supertranslation hair on Black Holes, *J. High Energy Phys.* **05**, 161.

Helliwell, T.M. and Konkowski, D.A. (1990). Electromagnetic fields in Khan–Penrose space-time, *Phys. Rev. D* **41**, 2507.

Herrera, L. and Santos, N.O. (2005). Cylindrical collapse and gravitational waves, *Class. Quantum Grav.* **22**, 2407.

Hill, C.T. Schramm, D.N. and Fry, J.N. (1989). Cosmological structure formation from soft topological defects, *Comm. Nucl. Part. Phys.* **19**, 25.

Hirschmann, E.W. Wang, A. and Wu, Y. (2004). Collapse of a scalar field in $2+1$ gravity, *Class. Quantum Grav.* **21**, 1791.

Hoenselaers, C. and Dietz, W. (1984). Solutions of Einstein's equations: techniques and results, in *Proceedings of the International Seminar on Exact Solutions of Einstein's Equations, in Retzbach*, Germany, November 14–18, 1983 (Springer-Verlag).

Hoenselaers, C. and Ernst, F.J. (1990). Matching pp-waves to the Kerr metric, *J. Math. Phys.* **31**, 144.

Hoenselaers, C. Kinnersley, W. and Xanthopoulos, B.C. (1979a). Generation of asymptotically flat space-times with any number of parameters, *Phys. Rev. Lett.* **42**, 481.

Hoenselaers, C. Kinnersley, W. and Xanthopoulos, B.C. (1979b). Symmetries of the stationary Einstein-Maxwell equations VI-transformations which generate asymptotically flat space-times with arbitrary multiple moments, *J. Math. Phys.* **20**, 2530.

Hogan, P.A. Barrabés, C. and Bressange, G.F. (1998). Colliding plane waves in Einstein–Maxwell theory, *Lett. Math. Phys.* **43**, 263.

Hogan, P.A. and Walsh, D.M. (2003). Collision of high frequency plane gravitational and electromagnetic waves, *Int. J. Mod. Phys.* **12**, 1459.

Holvorcem, P.R. Letelier, P.S. and Wang, A. (1995). The interaction of outgoing and ingoing spherically symmetric null fluids, *J. Math. Phys.* **36**, 3663.

Hu, N. (1983). *Proceedings of the Third Marcel Grossmann Meeting on General Relativity*, Shanghai, China (Science Press, Beijing).

Horava, H. and Witten, E. (1995). Heterotic and type I string dynamics from eleven dimensions, *Nucl. Phys. B* **460**, 506.

Horava, H. and Witten, E. (1996). Eleven-dimensional supergravity on a manifold with boundary, *Nucl. Phys. B* **475**, 94.

Ibañez, J. and Verdaguer, E. (1983). Soliton collision in general relativity, *Phys. Rev. Lett.* **51**, 1313.

Ibañez, J. and Verdaguer, E. (1986). Finite perturbations on Friedmann–Robertson–Walker models, *Astrophys. J.* **306**, 401.

Ipser, J. and Sikivie, P. (1984). Gravitational repulsive domain wall, *Phys. Rev. D* **30**, 712.

Ishihara, H. Takahashi, M. and Tomimatsu, A. (1988). Gravitational Faraday rotation induced by a Kerr block hole, *Phys. Rev. D* **38**, 472.

Islam, J.N. (1985). *Rotating Fields in General Relativity* (Cambridge University Press, Cambridge).

Israel, W. (1966). Singular hypersurfaces and thin shells in general relativity, *Il Nuovo Cimento B* **44**, 1.

Israel, W. (1967). Singular hypersurfaces and thin shells in general relativity, *Il Nuovo Cimento B* **48**, 463(E).

Ivanov, B.V. (1998). Colliding axisymmetric pp waves, *Phys. Rev. D* **57**, 3378.

Jantzen, R.T. (1980). Soliton solutions of the Einstein equations generated from cosmological solutions with additional symmetry, *Nuovo Cimento B* **59**, 287.

Jogia, S. and Griffiths, J.B. (1980). A Newman–Penrose formalism for space-tunes with torsion, *Gen. Relativ. Grav.* **12**, 597.

Jordan, P. Ehlers, J. and Kundt, W. (1960). *Akad. Wiss. Matnz. Abn. Math. -Nat. Kl., Jabrg* **2**.

Kachru, S. Schulz, M.B. and Trivedi, S. (2003). Moduli stabilization from fluxes in a simple IIB orientifold, *J. High Energy Phys.* **10**, 007.

Kerr, R.P. (1963). Gravitational field of a spinning mass as an example of alge-braically special metric, *Phys. Lett.* **11**, 237.

Khan, K.A. and Penrose, R. (1971). Scattering of two impulsive gravitational plane waves, *Nature (London)* **229**, 185.

Khorrami, M. and Mansouri, R. (1994). Cylindrically symmetric thin walls in general relativity, *J. Math. Phys.* **35**, 951.

Kibble, T.W.B. (1976). Topology of cosmic domains and strings, *J. Phys. A: Math. Gen.* **9**, 1387.

Kinnersley, W. (1969). Type D vacuum metrics, *J. Math. Phys.* **10**, 1195.

Kinnersley, W. and Walker, M. (1970). Uniformly accelerating charged mass in general relativity, *Phys. Rev. D* **2**, 1359.

Kitchingham, D.W. (1984). The use of generating techniques for space-times with two non-null commuting Killing vectors in vacuum and stiff perfect fluid cosmological models, *Class. Quantum Grav.* **1**, 677.

Kitchingham, D.W. (1986). The application of the homogeneous Hilbert problem of Hauser and Ernst to cosmological models with spatial axes of symmetry, *Class. Quantum Grav.* **3**, 133.

Konkowski, D.A. and Helliwell, F.M. (1989). Singularities in colliding gravitational plane wave space-times, *Class. Quantum Grav.* **6**, 1847.

Konkowski, D.A. and Helliwell, F.M. (1991). Stability of the quasiregular singularities in Bell–Szekeres spacetime, *Phys. Rev. D* **43**, 609.

Kramer, D. and Neugebauer, G. (1980). The superposition of two Kerr solutions, *Phys. Lett. A* **75**, 259.

Kramer, D. and Neugebauer, G. (1984). Backlund transformations in general relativity, in *Solutions of Einstein's Equations: Techniques and Results, Proceedings of the International Seminar on Exact Solutions of Einstein's Equations*, Retzbach, Germany, November 14–18, 1983, Vol. 2, eds. Hoenselaers, C. and Dietz, W. (Springer-Verlag).

Kramer, D. Stephani, H. Herlt, H. MacCallum, M. and Schmutzer, E. (1980). *Exact Solutions of Einstein's Field Equations* (Cambridge University Press, Cambridge).

Kundt, W. and Trumper, M. (1962). Beitrage zur theorte der gravitationas strahlungsfelder, *Alcad. Wiss. Lit. Mainz. Abhandl. Math.-Nat. Kl.* **12**.

Laguna-Castillo, P. and Matzner, R.A. (1986). Inflation and bubbles in general relativity, *Phys. Rev. D* **34**, 2913.

Lanczos, K. (1922). Bemerkung zur de Sttterschen Welt, *Phys. Z.* **23**, 539.

Lanczos, K. (1924). Flachenhafte verteliung der materie in der Einsteinschen gravitationstheorie, *Ann. Phys.* **74**, 518.

Landau, L.D. and Lifshitz, E.M. (1972). *The Classical Theory of Fields*, Fourth Revised English Edition, translated by M. Hamermesh (Pergamon Press).

Lasky, P.D. Thrane, E. Levin, Y. Blackman, J. and Chen, Y. (2016). Detecting gravitational wave memory with LIGO: implications of GW150914, *Phys. Rev. Lett.* **117**, 061102.

Letelier, P.S. (1986). Soliton solutions to the vacuum Einstein equations obtained from a non-diagonal seed solution, *J. Math. Phys.* **27**, 564.

Letelier, P.S. (1989). On soliton solutions to the vacuum Einstein equations obtained from a general seed solution, *Class. Quantum Grav.* **6**, 875.

Letelier, P.S. and Wang, A. (1993a). Collisions of cosmic walls, *Class. Quantum Grav.* **10**, L29.

Letelier, P.S. and Wang, A. (1993b). Spherically-symmetric thin shells in the Brans–Dicke theory of gravity, *Phys. Rev. D* **48**, 631.

Letelier, P.S. and Wang, A. (1993c). On the interaction of null fluids in cosmology, *Phys. Lett. A* **182**, 220.

Letelier, P.S. and Wang, A. (1994). Singularities formed by the focusing of cylindrical null fluids, *Phys. Rev. D* **49**, 5105.

Letelier, P.S. and Wang, A. (1995). Spacetime defects, *J. Math. Phys.* **36**, 3023.

Li, W. (1989). New families of colliding plane gravitational waves with collinear polarization, *Class. Quantum Grav.* **6**, 477.

Li, W. and Ernst, F.J. (1989). A family of electrovac colliding wave solutions of Einstein's equations, *J. Math. Phys.* **30**, 678.

Lichnerowicz, A. (1955). *Theories relatlvistes de la gravitation et de l'electromagnettsme* (Masson, Paris).

Lind, R.W. (1974). Shear-free, twisting Einstein–Maxwell metrics in the Newman–Penrose formalism, *Gen. Relativ. Grav.* **5**, 25.

Linde, A.D. (1984). The inflation universe, *Rep. Prog. Phys.* **47**, 925.

Lukas, A. Ovrut, B.A. Stelle, K.S. and Waldram, D. (1999). The universe as a domain wall, *Phys. Rev. D* **59**, 086001.

Lyth, D.H. and Liddle, A.D. (2009). *The Primordial Density Perturbation: Cosmology, Inflation, and the Origin of Structure* (Cambridge University Press, Cambridge).

Maartens, R. (2004). Brane-World Gravity, *Living Rev. Relativity* **7**, 7.

Maartens, R. and Koyama, K. (2010). Brane-World Gravity, *Living Rev. Relativity* **13**, 5.

MacCallum, M.A.H. (1984). Exact solutions in cosmology, in *Solutions of Einstein's Equations: Techniques and Results*, Proceedings of the International Seminar on Exact Solutions of Einstein's Equations, Retzbach, Germany, November 14–18, 1983, Vol. 2, eds. Hoenselaers, C. and Dietz, W. (Springer-Verlag).

Maggiore, M. (2008). *Gravitational Waves: Volume 1: Theory and Experiments* (Oxford University Press).

Matzner, R.A. and Tippler, F.J. (1984). Metaphysics of colliding self-gravitational plane waves, *Phys. Rev. D* **29**, 1575.

Misner, C.W. Thorne, K.S. and Wheeler, J.A. (1973). *Gravitation* (W.H. Freeman and Company).

Mizoguchi, S. (1996). Colliding wave solutions, duality, and diagonal embedding of general relativity in two-dimensional heterotic string theory, *Nucl. Phys. B* **461**, 155.

Nakao, K.-I. Harada, T. Kurita, Y. and Morisawa, Y. (2009). Relativistic gravitational collapse of a cylindrical shell of dust. II, *Prog. Theor. Phys.* **22**, 521.

Nakao, K.-I. Kurita, Y. Morisawa, Y. and Harada, T. (2007). Relativistic gravitational collapse of a cylindrical shell of dust, *Prog. Theor. Phys.* **117**, 75.

Neugebauer, G. (1979). Backlund transformations of axially symmetric stationary gravitational fields, *J. Phys. A: Math. Gen.* **12**, L67.

Neugebauer, G. (1980a). A general integral of the axially symmetric stationary Einstein equations, *J. Phys. A: Math. Gen.* **13**, L19.

Neugebauer, G. (1980b). Recursive calculation of axially symmetric stationary Einstein fields, *J. Phys. A: Math. Gen.* **13**, 1737.

Newman, E. and Penrose, R. (1962). An approach to gravitational radiation by a method of spin coefficients, *J. Math. Phys.* **3**, 566.

Newman, E. and Penrose, R. (1963). An approach to gravitational radiation by a method of spin coefficients, *J. Math. Phys.* **4**, 998(E).

Nogales, J.A.C. and Wang, A. (1998). Instability of cosmological event horizons of global non-static cosmic strings II: Perturbations of gravitational waves and massless scalar field, *Phys. Rev. D* **57**, 6089.

Nolan, B.C. (2000). Central singularity in spherical collapse, *Phys. Rev. D* **62**, 044015.

Nutku, Y. (1981). Comment on "collision of plane gravitational waves without singularities", *Phys. Rev. D* **24**, 1040.

Nutku, Y. and Halli, M. (1977). Colliding impulsive gravitational waves, *Phys. Rev. Lett.* **39**, 1379.

O'Brien, S. and Synge, J.E. (1952). Jump conditions at discontinuities in general relativity, *Comm. Dublin Inst. Adv. Stud.* **A9**.

Oliver, G. and Verdaguer, E. (1989). A family of inhomogeneous cosmological Einstein–Rosen metrics, *J. Math. Phys.* **30**, 442.

O'Raifeartaigh, L. (1972). *General Relativity*, Papers in Honour of J.L. Synge (Clarendon Press, Oxford).

Ori, A. (2000). Strength of curvature singularities, *Phys. Rev. D* **61**, 064016.

Paiva, F.M. and Wang, A. (1995). Geometry of planar domain walls, *Phys. Rev. D* **52**, 1281.

Palenta, S. and Meinel, M. (2017). A continuous Riemann-Hilbert problem for colliding plane gravitational waves, *Class. Quantum Grav.* **34**, 195011.

Pantaleo, X. (1979). *Relativity, Quanta and Cosmology*, Einstein 1879–1979 (New York, Johnson).

Papacostas, T. and Xanthopoulos, B.C. (1989). Collisions of gravitational and electromagnetic waves that do not develop curvature singularities, *J. Math. Phys.* **30**, 97.

Papapetrou, A. and Hamoui, A. (1968). Couches simples de mattere en relativite generale, *Ann. Inst. Henri Poincare* **IX**, 179.

Penrose, R. (1960). A spinor approach to general relativity, *Ann. Phys.* **10**, 171.

Penrose, R. (1963). Asymptotic properties of fields and space-times, *Phys. Rev. Lett.* **10**, 66.

Penorse, R. (1964). Conformal treatment of infinity, in *Relativity, Groups and Topology*, The 1963 Les Houches Lectures, eds. DeWitt, C. and DeWitt, B. (Gordon and Breach, New York), pp. 565–584.

Penrose, R. (1965). A remarkable property of plane waves in general relativity, *Rev. Mod. Phys.* **37**, 215.

Penrose, R. (1968). Structure of space-time, in *Battelle Rencontres, Lectures in Mathematical Physics*, eds. DeWitt, C. and Wheeler, J.A. (Benjamin, New York).

Penrose, R. (1969). Gravitational collapse: the role of general relativity, *Riv. Nuovo Cimento*, **1**, 252.

Penrose, R. (1972). The geometry of impulsive gravitational waves, in *General Relativity*, Papers in Honor of J.L Synge, ed. O'Raifeartaigh, L. (Oxford University Press, Oxford), pp. 101–115.

Pereira, P.R.C.T. and Wang, A. (2000a). Dynamic cylindrical shells with rotation in general relativity, *Gen. Relativ. Grav.* **32**, 2189.

Pereira, P.R.C.T. and Wang, A. (2000b). Gravitational collapse of counter-rotating cylindrical dust shells, *Phys. Rev. D* **62**, 124001.

Pereira, P.R.C.T. and Wang, A. (2002). Gravitational collapse of counter-rotating cylindrical dust shells, *Int. J. Mod. Phys. D* **11**, 561.

Petrov, A.Z. (1955). *New Methods in General Relativity* (Nauka, Moscow) (in Russian) [English edition of Petrov's book: Einstein Spaces, Pergamon Press, 1969].

Pirani, F. (1964). Introduction to gravitational radiation theory, in *Lectures in General Relativity*, Brandeis Summer Institute in Theoretical Physics, Vol. 1. eds. Deser, S. and Ford, K. (Prentice-Hall, Englewood Cliffs).

Piran, T. and Safier, P.N. (1985). A gravitational analogue of Faraday rotation, *Nature (London)* **318**, 271.

Piran, T. Safier, P.N. and Katz, J. (1986). Cylindrical gravitational waves with two degrees of freedom: an exact solution, *Phys. Rev. D* **34**, 331.

Piran, T. Safier, P.N. and Stark, R.F. (1985). General numerical solution of cylindrical gravitational waves, *Phys. Rev. D* **32**, 3101.

Pretorius, F. and East, W.E. (2018). Black hole formation from the collision of plane-fronted gravitational waves, *Phys. Rev. D* **98**, 084053.

Randall, L. and Sundrum, R. (1999a). Large mass hierarchy from a small extra dimension, *Phys. Rev. Lett.* **83**, 3370.

Randall, L. and Sundrum, R. (1999b). An alternative to compactification, *Phys. Rev. Lett.* **83**, 4690.

Ringeval, C. and Suyama, T. (2017). Stochastic gravitational waves from cosmic string loops in scaling, *J. Cosmol. Astropart. Phys.* **12**, 027.

Rocha, P. Chan, R. da Silva, M.F.A. and Wang, A. (2008a). Stable and "bounded excursion" gravastars, and black holes in Einstein's theory of gravity, *J. Cosmol. Astropart. Phys.* **11**, 010.

Rocha, P. Miguelote, A.Y. Chan, R. da Silva, M.F.A. Santos, N.O. and Wang, A. (2008b). Bounded excursion stable gravastars and black holes, *J. Cosmol. Astropart. Phys.* **06**, 025.

Rosen, N. (1937). Plane polarised waves in the general theory of relativity, *Phys. Z* **12**, 366.

Rosenbaum, M. Ryan, M. Urrutia, L.F. and Matzner, R.A. (1986). Colliding plane waves in $N = 1$ classical supergravity, *Phys. Rev. D* **34**, 409.

Roy, S. (2003). Accelerating cosmologies from M/string theory compactifications, *Phys. Lett. B* **567**, 322.

Ruffini, R. (1976). *Proceedings of the First Marcel Grossmann Meeting on General Relativity*, 7–12 July 1975 (North-Holland).

Sachs, R.K. (1960). Propagation laws for null and type III gravitational waves, *Phys. Z* **157**, 462.

Sachs, R.K. (1961). Gravitational waves in general relativity, IV: the outgoing radiation condition, *Proc. R. Soc. Lond. A* **264**, 309.

Sachs, R.K. (1962). Gravitational waves in general relativity, VIII: waves in asymptotically flat space-time, *Proc. R. Soc. Lond. A* **270**, 103.

Sachs, R.K. (1964). Gravitational radiation, in *Relativity, Groups and Topology*, eds. DeWitt, C. and DeWitt, B.S. (Gordon and Breach, New York).

Sanz, J.L. and Goicoechea, L.J. (1984). *Observational and Theoretical Aspects of Relativistic Astrophysics and Cosmology* (World Scientific Publishing Co.).

Sharma, P. Tziolas, A. Wang, A. and Wu, Z.-C. (2011). Spacetime Singularities in string and its low dimensional effective theory, *Int. J. Mod. Phys. A* **25**, 273.

Schmidt, H.-J. and Wang, A. (1993). Plan domain walls when coupled with the Brans-Dicke scalar field, *Phys. Rev. D* **47**, 4425.

Schwarz, P. (1997). Colliding axion–dilaton plane waves from black holes, *Phys. Rev. D* **56**, 7833.

Siegel, H.P. (1981). Evolving dust shells, *Phys. Rev. D* **23**, 2835.

Singh, P. and Griffiths, J.B. (1990). The application of spin coefficient techniques in the vacuum quadratic Poincare gauge field theory, *Gen. Relativ. Grav.* **22**, 269.

Souriau, J.-M. (1973). Ondes et radiations gravitationnelles, *Colloq. Int. CNRS*, *Paris* **220**, 243.

Spergel, D.N. *et al.* (WMAP) (2007). Wilkinson microwave anisotropy probe (WMAP) three year results: implications for cosmology, *Astrophys. J. Suppl.* **170**, 377.

Stark, R.F. and Connors, P.A. (1977), Observational test for the existence of a rotating black hole in CygX-1, *Nature (London)* **266**, 429.

Stephani, H. Kramer, D. MacCallum, M. Hoenselaers, C. and Herlt, E. (2009). *Exact Solutions of Einstein's Field Equations*, Cambridge Monographs on Mathematical Physics (Cambridge University Press, Cambridge).

Stoyanov, Yu.G. (1979). Interaction of plane gravitational waves, *Sov. Phys. JETP.* **44**, 1051.

Strominger, A. (2017). Lectures on the Infrared structure of gravity and gauge theory, preprint, arXiv:1703.05448.

Synge, J.L. (1965). *Relativity: The General Theory* (North-Holland).

Szekeres, P. (1965). The gravitational compass, *J. Math. Phys.* **6**, 1387.

Szekeres, P. (1966). On the propagation of gravitational fields in matter, *J. Math. Phys.* **7**, 751.

Szekeres, P. (1970). Colliding gravitational waves, *Nature (London)* **228**, 1183.

Szekeres, P. (1972). Colliding plane gravitational waves, *J. Math. Phys.* **13**, 286.

Tabensky, R. and Taub, A.H. (1973). Plane symmetric self-gravitating fluids with pressure equal to energy density, *Commun. Math. Phys.* **29**, 61.

Talbot, C.J. (1969). Newman–Penrose approach to twisting degenerate metrics, *Commun. Math. Phys.* **13**, 45.

Taub, A.H. (1956). Isentropic hydrodynamics in plane symmetric space-times, *Phys. Rev.* **103**, 454.

Taub, A.H. (1975). Spatially homogenous universe, in *Proceedings of the First Marcel Grossmann Meeting on General Relativity*, ed. Ruffini, R. (North-Holland).

Taub, A.H. (1980). Space-times with distribution valued curvature tensors, *J. Math. Phys.* **21**, 1423.

Taub, A.H. (1983). General relativistic hydrodynamic wave propagation, in *Proceedings of the Third Marcel Grossmann Meeting On General Relativity*, August 30–September 1, 1982, Shanghai, China, ed. Hu, N. (Science Press, Beijing).

Taub, A.H. (1988a). On the collision of planar impulsive gravitational waves, *J. Math. Phys.* **29**, 690.

Taub, A.H. (1988b). Collision of impulsive gravitational waves following dust clouds, *J. Math. Phys.* **29**, 690.

Taub, A.H. (1990). Interaction of null dust clouds fronted by impulsive plane gravitational waves, *J. Math. Phys.* **31**, 664.

Thorne, K.S. (1967). The general relativistic theory of stellar structure and dynamics, in *High-Energy Astrophysics*, vol. Ill, eds. DeWitt, C. Schatzman, E. and Veron, P. (Gordon and Breach, New York), p. 414.

Thorne, K.S. (1972). *In Magic without Magic: John Archibald Wheeler*, ed. J. Klauder (Freeman, San Francisco).

Thorne, K.S. (1992). Gravitational-wave bursts with memory: the Christodoulou effect, *Phys. Rev. D* **45**, 520.

Tippler, F.J. (1980). Singularities from colliding plane gravitational waves, *Phys. Rev. D* **22**, 2929.

Tomita, K. (1998). Plane-symmetric vacuum solutions with null singularities for inhomogeneous models and colliding gravitational waves, preprint, arXiv:gr-qc/9807033.

Townsend, P.K. and Wohlfarth, M.N.R. (2003). Accelerating cosmologies from compactification, *Phys. Rev. Lett.* **91**, 061302.

Tsoubelis, D. (1976). Homogeneous relativistic cosmological models of Bianchi type IV, Ph.D. dissertation, Graduate Faculty in Physics Department, The City University of New York.

Tsoubelis, D. (1989a). Cylindrical shells of cosmic strings, *Class. Quantum Grav.* **6**, 101.

Tsoubelis, D. (1989b). Plane gravitational waves colliding with shells of null dust, *Class. Quantum Grav.* **6**, L117.

Tsoubelis, D. and Wang, A. (1989). Asymmetric collision of gravitational plane waves: a new class of exact solutions, *Gen. Relativ. Grav.* **21**, 807.

Tsoubelis, D. and Wang, A. (1990). Impulsive shells of null dust colliding with gravitational plane waves, *Gen. Relativ. Grav.* **22**, 1091.

Tsoubelis, D. and Wang, A. (1991). On the gravitational interaction of plane symmetric null dust clouds, *J. Math. Phys.* **31**, 1017.

Tsoubelis, D. and Wang, A. (1992). Head-on collision of gravitational plane waves with non-collinear polarization: a new class of analytical models, *J. Math. Phys.* **33**, 1054.

Tziolas, A. and Wang, A. (2008). Colliding branes and formation of spacetime singularities, *Phys. Lett. B* **661**, 5.

Tziolas, A. Wang, A. and Wu, Z.-C. (2009). Colliding branes and formation of spacetime singularities in string theory, *J. High Energy Phys.* **04**, 038.

Verdaguer, E. (1984). Solitons and the generation of new cosmological solutions, in *Observational and Theoretical Aspects of Relativistic Astrophysics and Cosmology*, eds. Sanz, J.L. and Goicoechea, L.J. (World Scientific, Singapore).

Verdaguer, E. (1987). Plane waves and soliton solutions, *Il Nuovo. Cimento B* **100**, 787.

Verdaguer, E. (1993). Soliton solutions in spacetimes with two spacelike killing fields, *Phys. Rep.* **229**, 1.

Weyl, H. (1917). Zur gravitationstheorie, *Ann. Phys.* **54**, 117.

Vladimirov, V.S. (1979). *Generalized Functions in Mathematical Physics* (MIR, Moscow).

Vilenkin, A. (1981). Gravitational field of vacuum domain walls and strings, *Phys. Rev. D* **23**, 852.

Vilenkin, A. (1985). Cosmic strings and domain walls, *Phys. Rep.* **121**, 263.

Vilenkin, A. and Shellard, E.P.S. (2000). *Cosmic Strings and Other Topological Defects* (Cambridge University Press, Cambridge).

Wainwright, J. Ince, W.C.W. and Marshman, B.J. (1979). Spatially homogeneous and inhomogeneous cosmologies with equation of state, *Gen. Relativ. Grav.* **10**, 259.

Wald, R.M. (1984). *General Relativity* (University of Chicago, New York).

Wands, D. (2006). Brane-world cosmology, preprint, arXiv:gr-qc/0601078.

Wang, A. (1991a). The effect of polarization of colliding plane gravitational waves on focusing singularities, *J. Mod. Phys. A* **6**, 2273.

Wang, A. (1991b). Gravitational Faraday rotation induced from interacting gravitational plane waves, *Phys. Rev. D* **44**, 1120.

Wang, A. (1991c). Plane walls interacting with gravitational and matter fields, *J. Math. Phys.* **32**, 274.

Wang, A. (1991d). Massless particle radiation of Vilenkin's planar domain walls, *Phys. Lett. B* **264**, 2863.

Wang, A. (1991e). Planar domain walls emitting and absorbing electromagnetic radiation, *Phys. Rev. D* **44**, 1705.

Wang, A. (1991f). *Interacting Gravitational, Electromagnetic, Neutrino and Other Waves in the Context of Einstein's General Theory of Relativity* (Iaonnina University, Greece).

Wang, A. (1992a). Gravitational interaction of plane gravitational waves and matter shells, *J. Math. Phys.* **33**, 1065.

Wang, A. (1992b). Planar domain walls coupled with pure radiation fields, *Int. J. Mod. Phys. A* **7**, 4521.

Wang, A. (1992c). Singularities of the space-time of a plane domain wall when coupled with a massless scalar field, *Phys. Lett. B* **277**, 49.

Wang, A. (1992d). Transparency of domain walls to electromagnetic and neutrino waves: an exact solution, *Mod. Phys. Lett. A* **7**, 835.

Wang, A. (1992e). Dynamics of plane-symmetric thin walls in general relativity, *Phys. Rev. D* **45**, 3534.

Wang, A. (1992f). On the interaction of bubbles with gravitational and matter fields, *Mod. Phys. Lett. A* **7**, 1779.

Wang, A. (1993). Non-trivial interaction of plane domain walls with scalar fields, *Phys. Rev. D* **48**, 2591.

Wang, A. (1994). Gravitational collapse of thick domain walls: an analytic model, *Mod. Phys. Lett. A* **9**, 3605.

Wang, A. (2000). On "Comments on collapsing null shells of matter", *J. Math. Phys.* **41**, 8354.

Wang, A. (2001). Critical phenomena in gravitational collapse: The studies so far, *Braz. J. Phys.* **31**, 188.

Wang A. (2003). Critical collapse of cylindrically symmetric scalar field in four-dimensional Einstein's theory of gravity, *Phys. Rev. D* **68**, 064006.

Wang, A. (2010). Orbifold branes in string/M-theory and their cosmological applications, preprint, arXiv:1003.4991.

Wang, A. and de Oliveira, H.P. (1997). Critical phenomena of collapsing massless scalar wave packets, *Phys. Rev. D* **56**, 753.

Wang, A. and Letelier, P.S. (1995a). Local and global structure of a plane domain wall spacetime, *Phys. Rev. D* **51**, R6612.

Wang, A. and Letelier, P.S. (1995b). Domain wall spacetime: instability of cosmological event and Cauchy horizons, *Phys. Rev. D* **52**, 1800.

Wang, A. and Letelier, P.S. (1995c). Dynamical wormholes and energy conditions, *Prog. Theor. Phys.* **94**, L137.

Wang, A. and Nogales, J.A.C. (1997). Instability of cosmological event horizons of global non-static cosmic strings, *Phys. Rev. D* **56**, 6217.

Wang, A. and Santos, N.O. (1996). Gravitational and particle radiation from cosmic strings, *Class. Quantum Grav.* **13**, 715.

Wang, A. and Santos, N.O. (2008). The cosmological constant in the brane world of string theory on S^1/Z_2, *Phys. Lett. B* **669**, 127.

Wang, A. and Santos, N.O. (2010). The hierarchy problem, radion mass, localization of gravity and 4D effective Newtonian potential in string theory on S_1/Z_2, *Int. J. Mod. Phys. A* **25**, 1661.

Wang, T.-Z. Fier, J. Li, B. Lv, G.-L. Wang, Z.-J. Wu, Y. and Wang, A. (2018). Singularities of plane gravitational waves and their memory effects, preprint, arXiv:1807.09397 [*Gen. Relativ. Grav.* in press (2020)].

Whitham, G.B. (1974). *Linear and Nonlinear Waves* (Wiley, New York).

Weinberg, S. (1972). *Gravitation and Cosmology, Principles and Applications of the General Theory of Relativity* (Wiley, New York).

Westenholz, C. (1981). *Differential Forms in Mathematical Physics* (North-Holland, New York).

Weyl, H. (1917). Zur Gravitationstheorie, *Ann. Phys.* **54**, 117.

Weyl, H. (1922). *Space–Time–Matter*, translated by Brose, H. L. (Methuen, London).

Witten, L. (1962). *Gravitation: An Introduction on Current Research* (Wiley, New York).

Wohlfarth, M.N.R. (2003). Accelerating cosmologies and a phase transition in M-theory, *Phys. Lett. B* **563**, 1.

Wu, Q. Gong, Y.-G. and Wang, A. (2009). Brane cosmology in the Horava–Witten heterotic M-theory on S^1/Z_2, *J. Cosmol. Astropart. Phys.* **06**, 015.

Wu, Q. Vo, P. Santos, N.O. and Wang, A. (2008). Late transient acceleration of the universe in string theory on S^1/Z_2, *J. Cosmol. Astropart. Phys.* **09**, 004.

Wu, Y. da Silva, M.F.A. Santos, N.O. and Wang, A. (2003). Topological charged black holes in high dimensional spacetimes and their formation from gravitational collapse of a type II fluid, *Phys. Rev. D* **68**, 084012.

Xanthopoulos, B.C. (1979). Multiple moments in general relativity, *J. Phys. A: Math. Gen.* **12**, 1025.

Xanthopoulos, B.C. (1981). Exterior spacetimes for rotating stars, *J. Math. Phys.* **22**, 1254.

Xanthopoulos, B.C. (1986a). Rotating cosmic string, *Phys. Lett. B* **178**, 163.

Xanthopoulos, B.C. (1986b). Cylindrical waves and cosmic strings of Petrov type D, *Phys. Rev. D* **34**, 3608.

Xanthopoulos, B.C. (1986c). The initial value problem for colliding gravitational and hydrodynamic waves, *J. Math. Phys.* **27**, 2129.

Xanthopoulos, B.C. (1987). Cosmic strings coupled with gravitational and electromagnetic waves, *Phys. Rev. D* **35**, 3713.

Yurtsever, U. (1987). Instability of Killing-Cauchy horizons in plane-symmetric space-times, *Phys. Rev. D* **36**, 1662.

Yurtsever, U. (1988a). Structure of singularities produced by colliding plane gravitational waves, *Phys. Rev. D* **38**, 1706.

Yurtsever, U. (1988b). New family of exact solutions for colliding plane gravitational waves, *Phys. Rev. D* **37**, 2790.

Yurtsever, U. (1989). Singularities and horizons in the collisions of gravitational waves, *Phys. Rev. D* **40**, 329.

Zel'dovich, Ya.B. (1980). Cosmological fluctuations produced near a singularity, *Mon. Not. R. Astron. Soc.* **192**, 663.

Zel'dovich, Ya.B. Kobzarev, I.Yu. and Okun, L.B. (1976). Cosmological consequences of a spontaneous breakdown of a discrete symmetry, *Sov. Phys. JETP* **40**, 1.

Zeldovitch, Ya.B. and Polnarev, A.G. (1974). Radiation of gravitational waves by a cluster of superdense stars, *Sov. Astron.* **18**, 17.

Zemanian, A.H. (1987). *Distribution Theory and Transform Analysis: An Introduction to Generalized Functions, with Applications* (Dover Publications, New York).

Zhang, P.-M. Duval, C. Gibbons, G.W. and Horvathy, P.A. (2017). The memory effect for plane gravitational waves, *Phys. Lett. B* **772**, 743.

Zhang, P.-M. Duval, C. Gibbons, G.W. and Horvathy, P.A. (2018). Velocity memory effect for polarized gravitational waves, *J. Cosmol. Astropart. Phys.* **05**, 030.

Index

Printed in the United States
By Bookmasters